Build Your Own PC on a Budget

Build Your Own PC on a Budget

A DIY Guide for Hobbyists and Gamers

John Paul Mueller

New York   Chicago   San Francisco
Athens   London   Madrid
Mexico City   Milan   New Delhi
Singapore   Sydney   Toronto

Library of Congress Cataloging-in-Publication Data

Mueller, John, date.
 Build your own PC on a budget : a DIY guide for hobbyists and gamers / John Paul Mueller.
 —First [edition].
    p. cm.
 ISBN 978-0-07-184237-2 (paperback)
 1. Microcomputers—Design and construction—Amateurs' manuals. I. Title.
TK9969.M83 2015
621.39'16—dc23   2015033244

McGraw-Hill Education books are available at special quantity discounts to use as premiums and sales promotions or for use in corporate training programs. To contact a representative, please visit the Contact Us page at www.mhprofessional.com.

Build Your Own PC on a Budget: A DIY Guide for Hobbyists and Gamers

Copyright © 2016 by McGraw-Hill Education. All rights reserved. Printed in the United States of America. Except as permitted under the United States Copyright Act of 1976, no part of this publication may be reproduced or distributed in any form or by any means, or stored in a database or retrieval system, without the prior written permission of the publisher.

1 2 3 4 5 6 7 8 9 0 DOC/DOC 1 2 1 0 9 8 7 6 5

ISBN   978-0-07-184237-2
MHID   0-07-184237-3

This book is printed on acid-free paper.

Sponsoring Editor
 Michael McCabe

Editing Supervisor
 Stephen M. Smith

Production Supervisor
 Pamela A. Pelton

Acquisitions Coordinator
 Lauren Rogers

Project Manager
 Dheeraj Chahal, MPS Limited

Copy Editor
 Md. Taiyab Khan, MPS Limited

Proofreader
 Alekha Jena, MPS Limited

Indexer
 Alexandra Nickerson

Art Director, Cover
 Jeff Weeks

Composition
 MPS Limited

McGraw-Hill Education, the McGraw-Hill Education logo, TAB, and related trade dress are trademarks or registered trademarks of McGraw-Hill Education and/or its affiliates in the United States and other countries and may not be used without written permission. All other trademarks are the property of their respective owners. McGraw-Hill Education is not associated with any product or vendor mentioned in this book.

Information contained in this work has been obtained by McGraw-Hill Education from sources believed to be reliable. However, neither McGraw-Hill Education nor its authors guarantee the accuracy or completeness of any information published herein, and neither McGraw-Hill Education nor its authors shall be responsible for any errors, omissions, or damages arising out of use of this information. This work is published with the understanding that McGraw-Hill Education and its authors are supplying information but are not attempting to render engineering or other professional services. If such services are required, the assistance of an appropriate professional should be sought.

This book is dedicated to my agent, Matt Wagner. Over the years, Matt has inspired me to expand my horizons and have a bit of fun while doing it. One of the reasons I've written so many different kinds of books is because Matt is always there to encourage and to present some scathingly brilliant ideas.

About the Author

John Paul Mueller is a freelance author and technical editor. He has writing in his blood, having produced 98 books and more than 600 articles to date. The topics range from networking to home security and from database management to heads-down programming. Some of John's current books include one on Python for beginners and another about MATLAB. His home security book discusses all sorts of hardware issues, including the installation of various sensors. John's technical editing skills have helped more than 63 authors refine the content of their manuscripts. He has provided technical editing services to both *Data Based Advisor* and *Coast Compute* magazines. During his time at Cubic Corporation, John was exposed to reliability engineering and has had an interest in probability ever since. Be sure to read his blog at http://blog.johnmuellerbooks.com/.

When John isn't working at the computer, you can find him outside in the garden, cutting wood, or generally enjoying nature. He also likes making wine, baking cookies, and knitting. When not occupied with anything else, John makes glycerin soap and candles, which come in handy for gift baskets. You can reach him on the Internet at John@JohnMuellerBooks.com. John is also setting up a website at http://www.johnmuellerbooks.com/. Feel free to take a look and make suggestions on how he can improve it.

Contents

Acknowledgments xxi

Introduction xxiii

Part I—Developing a PC Plan

1 Defining What You Want 3
    Considering Why Off-the-Shelf Systems Aren't Appealing 3
        Dealing with Standard Off-the-Shelf Systems 4
        Dealing with Customizable Off-the-Shelf Systems 6
        Employing the Bare-Bones Kits 7
    Writing Down Goals for Your System 8
    Creating a Budget 9
    Identifying the Parts You Have Now 11
        Defining What Reusable Means 11
        Obtaining Documentation for Existing Parts 12
        Ensuring the Part Is Actually Usable 12

2 Introducing the Major Parts 15
    Opening the Case 16
    Understanding the Role of the Motherboard 18
    Providing Power to the System 21

	Considering the Add-on Boards	23
	Working with Display Adapters	23
	Hearing with Sound Cards	25
	Providing Interfaces Using Host Adapters	26
	Controlling Access Using Security Device Cards	26
	Considering Other Add-on Boards	27
	Deciding on Storage Devices	27
	Connecting with Cables	29
	Keeping Things Cool	30
	Focusing on External Connectivity	33
	Dealing with Network Cards	33
	Dealing with Port Cards	34
3	**Considering the Vendors**	**35**
	Getting on Vendor Sites	36
	Performing Quick Comparisons	37
	Obtaining Part Information in the Correct Order	38
	Performing Apples-to-Apples Comparisons	40
	Understanding the Value of Testing	40
	Performing Comparisons Efficiently	41
	Finding Reliable Reviews	42
	Getting Good Professional Reviews	43
	Understanding the Value of Points	43
	Using the Good Review Checklist	44
	Understanding the Compatibility Pitfalls	45
	Considering Problems with Standards Adherence	45
	Considering Problems with Feature Extensions	46
	Working Through the Odd Bad Part	46
	Dealing with Compatibility Issues	47
	Reading Between the Lines	47
4	**Getting What You Need**	**49**
	Creating and Verifying a Purchase List	50
	Dealing with Specifics	50
	Considering Reputation	50

Contents

Performing Comparison Shopping	51
Verifying the Warranty	51
Checking the Return Policy	52
Making the Purchases	52
Allowing Scripts	53
Getting Your Information Together	53
Completing Your Purchase	55
Verifying the Package Contents	56
Checking the Boxes	56
Checking the Individual Parts Box Content	57
Making a List of Extras	58
Getting the Extras You Need	59
Ensuring the Documentation Is Complete	60
Inventorying the Required Tools	61
Obtaining a Computer Toolkit	61
Getting Individual Tools	63
Setting Up a Work Area	64
Selecting a Worktable	64
Ensuring You Have Enough Light	64
Checking the Outlets	65
Keeping Your Work Area Clean	65

Part II—Building the Hardware

5 Adding RAM and Processor ... 69

Understanding Static Electricity	69
Verifying the Processor and RAM Positions	70
Adding Cooling to the Processor	71
Understanding the Need for Thermal Paste	72
Working with Standard Processor Cooling	73
Understanding Liquid Processor Cooling	74
Inserting the Processor	75
Orienting the Processor	75
Locking the Processor Down	76

Installing the RAM	...	79
Looking at the RAM Sockets	...	79
Inserting and Securing the RAM	...	79

6 Installing the Motherboard — 81

Configuring the Case	...	81
Installing a Power Supply	...	82
Identifying the Power Plugs and Sockets	...	84
Setting Up the Motherboard	...	86
Ensuring the Motherboard Is in Place	...	87
Installing the Standoffs	...	87
Positioning the Motherboard	...	89
Screwing the Motherboard in Place	...	90
Connecting the Case Features to the Motherboard	...	91
Identifying the Pins on the Motherboard	...	92
Connecting the Case Plugs	...	94
Testing Your Initial Setup	...	95

7 Providing Video — 97

Understanding the Video Basics	...	97
Understanding How Things Work	...	98
Using a Special Motherboard Socket	...	98
Finding the Special Power Supply Connection	...	99
Viewing the Back of the Card	...	99
Installing the Video Card Correctly	...	101
Connecting Any Required Cables	...	104
Considering CrossFire and SLI Configuration Needs	...	105
Configuring the Motherboard	...	106
Using the Correct Slot	...	107
Making the Required Power Supply Connections	...	108
Making Connections Between Cards	...	108
Considering TV Tuner Configuration Needs	...	108

Contents   xiii

       Connecting Devices 109
       Performing a Quick Video Test 112

8 Mounting Permanent Storage **115**
       Understanding Permanent Storage Basics 116
            Defining the Form Factors 117
            Considering the Power Cable 118
            Considering the Data Cable 120
            Working with Drive Size Adapters 122
       Installing a SATA Drive 124
       Installing an Optical Drive 126
       Working with Solid-State Drives 127
       Using USB Storage 128
       Working with External Drives 130
       Considering Other Permanent Storage Options 130
            Considering the SCSI Option 131
            Understanding RAID 131

9 Attaching Auxiliary Devices **135**
       Choosing Keyboard and Other Input
            Devices Carefully 136
       Connecting the Keyboard 137
            Considering the Advantages of Wired Keyboards 137
            Working with Wired Keyboards 138
            Working with Wireless Keyboards 139
       Connecting the Mouse or Trackball 140
            Considering the Trackball Difference 141
            Working with a Wired Mouse or Trackball 142
            Working with a Wireless Mouse or Trackball 143
       Testing the Basic Setup 144
       Adding a Printer 145
       Working with Webcams 146
       Providing Other Device Connections 146

Part III—Considering Networks

10 Installing a LAN 151
Understanding LAN Basics 151
Considering the LAN Hardware 152
Creating the Physical Connection 153
Considering the LAN Software 154
Seeing the Other Connections 155
Configuring the Motherboard LAN 156
Dealing with Jumpers 156
Performing the System Configuration 157
Considering Multiple-LAN Motherboards 157
Installing a Separate LAN Card 158
Using External LAN Solutions 159
Understanding WLAN Essentials 160
Defining the WLAN 160
Delving into Wi-Fi 161
Delving into WiMAX 162
Considering LAN Security 163
Developing a Security Plan 163
Creating Useful Passwords 165
Defining the Hazards of WLAN 166

11 Connecting to the Internet 167
Configuring Multiple LANs 167
Considering the Need for a Router 168
Connecting to the LANs 169
Determining Your Internet Connection Type 169
Connecting a MODEM 172
Defining the Purpose of a MODEM 172
Obtaining the Correct MODEM 172
Configuring the MODEM 173
Connecting to the MODEM 174
Using a Test System to Check the Connection 174

Using Alternative Connectivity Options 174
Considering Internet Security 175
   Defining the Internet Difference 175
   Locking Your System Down 176
   Locking the Browser Down 176
   Relying on Firewalls 176
   Using Other Security Measures 177

12 Accessing Wireless Devices 179
Understanding the Common Wireless Standards 180
Ensuring You Have the Correct Wireless Support 180
Configuring Common Computer Devices 185
Configuring Alternative Devices 186
Considering Wireless Device Security 187
   Defining the Types of Intrusions 188
   Protecting the Communication Channel 189
   Protecting the Network 190
   Understanding the Role of Wireless Access Points 191

Part IV—Installing the Software

13 Installing the Operating System 195
Choosing an Operating System 195
   Defining What You Want to Do 196
   Considering Version Issues 197
   Considering Longevity 198
   Exploring Alternative Solutions 199
Obtaining the Vendor Installation Instructions 200
   Locating the Windows Instructions 201
   Choosing a Linux Flavor 201
   Hacking OS X 203
Ensuring You Have All the Details 203
   Getting a Licensed Copy 203
   Obtaining Manuals 204

	Performing Backups	204
	Downloading Device Drivers	205
Performing the Installation		205

14 Accessing the Devices ... 207

Understanding the Operating System to Device Connection		208
	Considering the Use of Low-Level Resources	208
	Avoiding Device Conflicts	210
	Circumventing Application Conflicts	211
Relying on Operating System Drivers		211
	Defining the Advantages of Automatic Access	212
	Considering the Missing Device	212
Obtaining and Using Device Vendor Drivers		212
	Considering the Advantages of Vendor Drivers	213
	Keeping Vendor Drivers Updated	215
Installing the Device Drivers		218
	Documenting Driver Settings	218
	Performing a System Backup	218
	Removing Existing Drivers	219
	Performing Required System Reboots	220
	Installing the New Driver	220
Overcoming Driver-Specific Problems		221
	Ensuring Device Connectivity	222
	Disabling the Driver Temporarily	222
	Verifying Driver Settings	223
	Considering the Device Driver Mismatch	224

15 Choosing Applications ... 225

Matching Applications to Your System		226
Verifying the Hardware Requirements		226
	Validating the Platform Requirements	227
	Checking Version Information	228
	Considering Connectivity Requirements	228
Obtaining Application Reviews		229

Installing the Application 230
    Avoiding Installation Complexities 230
    Using an Installation Program 231
    Unpacking the Application 232
    Working with Web-Based Applications 232
Correcting Application Installation Problems 233
    Dealing with Mismatches 233
    Finding Missing Components 234
    Getting Peer Help 234
    Resetting an Installation 235
Removing Errant and Unused Applications 236

Part V—Performing Maintenance

16 Maintaining the Hardware 239

Cleaning the Outside 240
    Vacuuming the Louvers and Cables 240
    Wiping the Case 240
    Checking for Damage 241
Cleaning the Inside 242
    Opening the Case 242
    Spraying the Dust Out 244
    Cleaning the Filters 245
Inspecting Your Hardware 245
    Checking the Add-on Boards 246
    Checking the Storage Devices 246
    Checking the Cables 248
Keeping Things Cool 248
    Spinning the Fan Blades 248
    Performing a Powered Test 249
Obtaining Spare Parts 250
    Creating a Parts List 250
    Shopping for Older Parts 251
    Getting Updated Parts 251

	Maintaining Essential Spares	252
	Creating a Maintenance Schedule	252
17	**Managing the Software**	**255**
	Performing Required Updates	255
	Setting Software to Check Updates Automatically	256
	Validating the Update	257
	Downloading and Installing the Update	258
	Testing the Update	259
	Keeping Things Secure	259
	Starting with Passwords	260
	Installing the Correct Background Applications	260
	Applying Security Updates Quickly	261
	Reading Those Installation and Update Screens	261
	Dealing with System Slowdowns	263
	Removing Unused Applications	264
	Cleaning the Hard Drive	264
	Cleaning Up the Operating System	265
	Keeping Data Safe	266
	Performing Backups	266
	Keeping Backups Off-Site	267
	Encrypting Data as Needed	267
	Overcoming Disasters	267
	Checking Drivers, Services, and Agents	268
	Testing Errant Hardware	269
	Restoring a Backup	269
18	**Preparing for Updates**	**271**
	Maintaining a Wish List	272
	Building with Expansion in Mind	273
	Developing a Processor Plan	273
	Upgrading Your RAM	275
	Upgrading a Display Adapter	275
	Adding a Display Adapter	276

Increasing Your Storage	277
Obtaining Special Add-ons	277
Considering the Role of Software	278
Obtaining and Installing Hardware Updates	279
Knowing When to Retire Your Old System	280
Index	**281**

Acknowledgments

Thanks to my wife, Rebecca. Even though she is gone now, her spirit is in every book I write, in every word that appears on the page. She believed in me when no one else would.

I extend a special thank you to Pegg Conderman for shooting all the photographs you see in the book. She did an amazing job and worked hard to get just the right shot. The book wouldn't be nearly as good without her diligent efforts.

Matt Wagner, my agent, deserves credit for helping me get the contract in the first place and taking care of all the details that most authors don't really consider. I always appreciate his assistance. It's good to know that someone wants to help.

A number of people read all or part of this book to help me refine the approach, test scripts, and generally provide input that all readers wish they could have. These unpaid volunteers helped in ways too numerous to mention here. I especially appreciate the efforts of Russ Mullen, Eva Beattie, Glenn A. Russell, and Austin Jonas, who provided general input, read the entire book, and selflessly devoted themselves to this project.

Finally, I would like to thank Michael McCabe and the rest of the editorial and production staff.

Introduction

Building your own PC might seem like a truly horrid idea in a day when computers have become commodities and you can get a decent setup for a song. However, that's not what building your own PC is all about. When you build your own PC, you decide what capabilities it provides, what the quality of the components will be, precisely how the system is configured, and how it looks when you turn it on. A truly custom PC is a work of art, a wonder, an amazement, and something others will envy when they see it. Most importantly, your custom-built PC will do precisely what you want it to do, when you want it to do it. *Build Your Own PC on a Budget* is your ticket to an adventure that many have wanted to take, but few have actually experienced. In this book, you see a path to the exotic wonders of creating a piece of hardware that everyone uses, but few understand.

About This Book

Build Your Own PC on a Budget is designed to help you create a system that will last a long time, yet outperform anything you could purchase in a store, all for a price much lower than the boutique stores that generally sell these high-end systems. This book takes you through a process beginning with selecting just the right components and ending with a functional system that's ready for use. In fact, it goes further than any other book on the topic because it helps you maintain your investment after you build it. This book takes you through the following areas:

- Part I—Developing a PC Plan: You begin by deciding precisely what sort of system you want to build. The chapters help you understand what sorts of hardware work best and how to avoid potential problems when choosing your hardware. You get introduced to vendors and discover how to speak the way they do so you can get the best possible deal. When you

finish this part of the book, you know precisely what you want to build and have the parts on hand to perform the task.
- Part II—Building the Hardware: It's not possible to just stick the parts into the case in any order and have a functional PC when you're done. A PC is built in a specific order and using specific skills to ensure the best result. The chapters in this part emphasize safe building techniques and show you how to test the PC you're building each step of the way so that you end up with a truly wonderful system that works precisely as you thought it would. This part of the book also discusses how to attach auxiliary and specialty devices, so they're ready for use when you install the required software later in the book.
- Part III—Considering Networks: Many operating systems today rely on a network connection for various purposes, such as activation and obtaining updates during the installation process. Consequently, it's important to have a network in place before you install anything to ensure you have the required access during the installation process. The chapters in this part help you avoid the major pitfalls of LAN setup and configuration so that you don't end up fighting with your network while installing your operating system and applications.
- Part IV—Installing the Software: The chapters in this part help you install an operating system and applications. It places a special emphasis on device drivers because many of the special devices that people attach to custom-built machines have special device drivers and associated software as well. When you get through this part of the book, you can begin using your PC if you want.
- Part V—Performing Maintenance: A lot of people think that you can build a PC and then never clean it. They're mystified when smoke begins to rise out of the vent holes one day, the PC makes an odd sort of groaning sound, and then all their data disappears because they never thought to back it up. In the pooled metal and plastic that was once their computer, their data lies, unreachable. Don't let this happen to you! This part of the book helps you maintain your system so that it performs well for a long time and actually protects the precious data you've spent so much time accumulating.

Icons Used in This Book

Sometimes it's important to make a point—to make text more important for some reason or another than the text around it. Icons help make the point in this book. Whenever you see an icon, you know the associated text is special for some reason. Here are the icons used in this book:

NOTE

A note describes something that's beyond the general discussion, but that you could find helpful or important. You should read notes with the idea that they provide ancillary information. Having the note icon makes the material easier to find when you discover that you really should have remembered the content that the note provided. Pay particular attention to online resources that notes point out because they often help you create a better system.

TIP

Most people want to know what makes the difference between a novice and an expert. It's the fact that an expert has experiential knowledge the novice lacks. Tips are experiential knowledge. They help turn you from a novice to an expert in just a few seconds. Whenever you see a tip, think about information that you wouldn't normally find in a book that follows all the rules. This information is based on how the real world works.

WARNING

Even new PCs blow up. In fact, they blow up quite a bit easier than older PCs do simply because they're new and not broken in yet. If you want to keep from totally trashing your system before you even get to use it the first time, make sure you pay particular attention to warnings. These bits of text help you understand that doing something could potentially hurt your system and turn it into so much junk.

What You Should Know

This book doesn't assume you've built a PC before. That's the whole reason for buying the book. However, you do have to have a few basic skills in order to get the most possible from the book. The first rule is that you need to follow instructions carefully and know the difference between a screwdriver and a plier. Having some idea of which tools are which is really handy. Yes, you do get some information on which tools you actually need, but having some idea of what they look like in real life is important.

This isn't a book on operating systems or applications. Yes, it does mention them as part of the building process. You also gain some special insights into how operating systems and applications tend to interact with custom systems.

However, in order to be successful in actually creating a working system, you need to know the basics of how to use the operating system and applications of your choice.

Use the Online Resources

This isn't the end of your experience. As people write me, I'll provide you with some additional blog posts for this book at http://blog.johnmuellerbooks.com/category/technical/build-your-own-pc-on-a-budget/. Make sure you check out my blog from time to time to get the latest information on building your own system. You can contact me at John@JohnMuellerBooks.com to get book-specific questions answered. I want to be sure that your PC building experience is the best that I can make it.

Build Your Own PC on a Budget

Part I
Developing a PC Plan

1
Defining What You Want

It's likely that you have some vision for your new computer system. You know that you want something faster, more reliable, and with better features than the system you looked at online or at the store the other day. However, the specifics of this new system might be a little fuzzy, which is a problem because you can't build a system until you know precisely what to build.

You need to consider just how you want to make your system faster and what you're willing to pay for reliability. If you're working with a lot of graphics, then it might be better to invest in a great video adapter, rather than in more RAM (don't worry too much if you don't understand these terms—you will find them described as the book progresses). Likewise, it's important to consider precisely what features you want to install. For example, if you want to attach a wireless camera for a security system, you need to think about the specifics of that camera. In short, you need some sort of blueprint (specification) for your ideal system.

The chapter helps you understand the purpose of defining the system you want and the techniques you can use to accomplish the task. This is a starting point for every system you want to build—at least, if you want to end up with that spiffy system you have in mind. Having a clear plan in mind also does things like keep you on schedule and on budget. So, even though this chapter doesn't help you actually put parts into a cabinet, it's an essential chapter.

Considering Why Off-the-Shelf Systems Aren't Appealing

Vendors who build off-the-shelf systems are focused on creating an inexpensive product for you, while maximizing profits for themselves. The profit margins for today's systems are razor thin and getting thinner all the time. In order to obtain the required parts, build the systems, and still make a profit, vendors have to make huge bulk purchases and shave every last penny they can off the cost of the system components and the building process. The margins are so thin that designers must often consider the price of items like screws.

There are a myriad of ways in which off-the-shelf systems are presented. However, you can classify them in one of three ways:

- An off-the-shelf system can be offered as a single entity, where you buy the device without making any changes to it whatsoever. Each device variation actually has a different stock number associated with it. Typically, you can add memory to these setups, but nothing else.
- Some vendors offer base systems where you add components from a standardized list. The number of permutations is small, but you do gain a little flexibility when using this approach.
- A bare-bones kit starts with an off-the-shelf system, but then you buy components that you want to add to it as a separate investment. You gain even more flexibility over the customizable system, but you still start with a set of components that the vendor has chosen for you and now you face the prospect of reduced vendor support and warranty coverage.

It's important to understand the consequences of choosing an off-the-shelf system. The following sections describe them in more detail.

Dealing with Standard Off-the-Shelf Systems

People sometimes wonder what is so bad about off-the-shelf systems. The fact is that they aren't bad options for people who have generalized computer needs. If all you plan to do is work with a word processor or browse the Internet, an off-the-shelf system is probably fine for your needs. Standard off-the-shelf systems come with a specific set of features as shown in Figure 1-1.

However, once you start thinking about gaming or special functionality, the off-the-shelf systems become less appealing. Remember that the main emphasis of off-the-shelf systems today is price—building a system for the lowest price possible regardless of what that means in other ways. In fact, here are some reasons that you want to build a custom system, rather than using off-the-shelf options:

- Expansion: Most off-the-shelf systems don't offer much, if anything, in the way of expansion potential. Adding new or updated components becomes a nightmare because the system is often designed to preclude changes.

Chapter 1: Defining What You Want

FIGURE 1-1 Most standard off-the-shelf systems come with a specific set of features.

- Documentation: It's hard to get documentation for your off-the-shelf system. Without good documentation, it's hard to know whether you can update a processor, add RAM, or determine what kind of drive updates to use. Changing the configuration to accommodate special components is also nearly impossible.
- Reliability: Vendors tend to use the least expensive components possible to keep prices down. This means that any sort of stress on the system, such as adding components, is likely to cause it to fail.
- Maintainability: An off-the-shelf system isn't designed for end-user fiddling, which means that when you open the case, you'll likely find that everything is secured in a manner that makes it hard to change anything. In fact, the mere act of opening the case often voids any warranty offered by the vendor.
- Performance: When you build your own system you can control the components you put into it, which means that you also control the system speed and other characteristics. A vendor has to build a system that performs well in all sorts of situations. You can build a system that performs exceptionally well in your particular situation—you don't need to consider anyone else's needs.

TIP

One of the biggest factors in creating a high-performance system on a budget is to choose the parts you use carefully. Using slower parts where you don't need the speed can help you save money for items where speed actually does matter. You can selectively cut corners and add enhancements to obtain precisely the system you need, rather than settle for someone else's idea of what you want.

Dealing with Customizable Off-the-Shelf Systems

Some vendors make it possible for you to build your own system using off-the-shelf building blocks. The idea is to offer some level of flexibility, while keeping the number of options low so that stocking them doesn't become a problem and so the vendor still gets the benefit of making bulk component purchases. In most cases, you start with a bare-bones case and then add components to it using a menu or wizard approach like the one shown in Figure 1-2.

In addition to making the system more flexible, using the customizable approach has fewer risks because the vendor ensures that the various components actually go together. When you build your own system, you must ensure that the

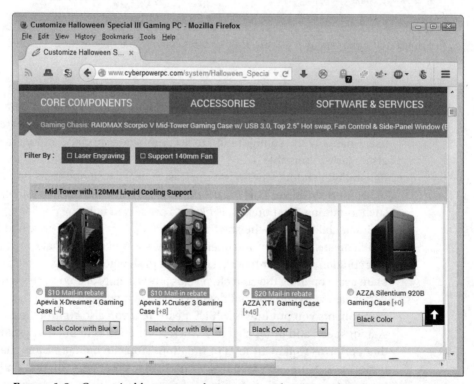

FIGURE 1-2 Customizable systems rely on a menu where you select the items you want in the case.

components actually work together. In some cases, this means making phone calls to the vendor and also reading online reviews from others who are custom building their systems. Unfortunately, a custom system still has many of the problems of a standard off-the-shelf system. For example, you may not get the graphics speed you expect when none of the choices include a high-performance display adapter.

WARNING

It might seem as if a customizable system would be expandable. However, this isn't always the case. A customizable system may use an off-sized motherboard or case that severely limits the expansion potential of the system (assuming there is any expansion capability at all). In addition, the vendor may not supply any documentation. You must ensure that every feature you need is added to the system when you buy it, not as an afterthought.

Employing the Bare-Bones Kits

When you build your own PC, you need to consider issues like the kind of processor you want and the sort of motherboard you need. A bare-bones kit, like the one shown in Figure 1-3, provides you with a case, motherboard, and power

FIGURE 1-3   Bare-bones systems provide you with the starting essentials you need for a system.

supply most of the time (some kits include additional features). You need to supply the other items needed to create a working PC. In this case, you must pay particular attention to the system specification to ensure you know what is and isn't supplied.

The benefit of a bare-bones system is that you have a starting point, but still have the flexibility needed to make some decisions. So, you get rid of more of the problems with a standard off-the-shelf system. For example, a bare-bones system is fully expandable (within the limits of the motherboard and case) by definition. In addition, because the vendor expects you to provide additional components, documentation is usually less of a problem.

Unfortunately, bare-bones systems can come with hidden issues. For example, if something goes wrong during the building process, the vendor may point the finger at you or the vendor of any component you add to the case. The warranty you thought you had for the bare-bones components may go up in smoke. Make sure you understand all the particulars of a bare-bones system before you get one.

Writing Down Goals for Your System

As described in the previous sections, off-the-shelf systems have a number of issues that are difficult or impossible to overcome if you want to experiment with your system, need to add any sort of specialized functionality, or simply need speed that an off-the-shelf system can't provide. Unfortunately, it's all too easy to become zombie-eyed after a few hours of staring at the ads online and completely forget what you were intending to do at the outset. First-time systems often end up looking like Frankenstein monsters and costing a small fortune to build. Of course, the way to avoid that problem is to define some goals for your system before you go online and look at anything.

The best way to define goals is to think about what you want to do with your system. For example, if you plan to perform graphics-intensive tasks, such as gaming, then you need a better-than-average display adapter. On the other hand, if you plan to build a security system or to experiment with various sorts of external devices, then the kind and placement of ports on your system become important.

It's also important to organize your goals. Make sure you address the needs of the highest priority goals first, and then work your way down the list. You might find that some goals fall into the category of future expansion because your budget won't support them now. Making tradeoffs is part of building a PC. You want to minimize the number of tradeoffs, but you need to realize that you may not be able to meet all your goals immediately.

Expansion is where the greatest potential of your system comes into play. When buying off-the-shelf you have to buy a whole new system every time your

Chapter 1: Defining What You Want

goals change. Building your own PC means that you can simply update one or two components as your goals change or as you have more money to make updates.

Take time to create your list of goals now, before you proceed any further into the chapter. You can summarize the goal creation process as:

1. Create a list of goals for your system.
2. Organize the goals in order of priority.
3. Consider which goals may have to be part of a future expansion and mark them as such.

Creating a Budget

Few people can afford the system they really want—at least, not at the outset. Some of the best systems out there were built a component at a time over an extended timeframe. Money comes into the picture pretty quickly when you start to figure out what you want in the way of a system. So, it's important to create a budget for your system. A basic system must have these components—each of which are purchased separately for the most part (don't worry about the weird terms in the following list—you will discover what they mean in future chapters):

- Case (starting price, $20)
- Power supply (starting price for a bare-bones 500-W supply, $15)
- Motherboard (starting price, $50)
- Processor (starting price for an Intel Core i5-3470S Quad-Core Processor, $196)
- Random access memory (RAM) (starting price for 4-GB, 1333-MHz, DDR3, $50)
- Display adapter (starting price, $40)
- Permanent storage (starting price for a 500-GB drive, $47)
- Network card (usually included with motherboard, but plan to spend a minimum of $8 if it isn't)

Just getting the items in the case is going to cost you $426 if you buy the least expensive components with any level of reliability at all. In addition to this cost, you need to add a monitor, mouse, keyboard, and some sort of sound output device, such as a speaker. So, it doesn't take long to figure out that you can't build an exact replica of the off-the-shelf systems out there for the same price that the vendor offers them to you at your local store. The reason you build a PC is because you want to have something that is fully documented, expandable, fixable, and flexible.

You can often buy refurbished parts for a fraction of the cost of new parts. Make sure you get a new part warranty when buying a refurbished part and test the part immediately upon receiving it if at all possible. Refurbished parts are parts that someone has sent back to the vendor with a problem in many cases (some refurbished parts are simply returns that didn't meet the original user's needs). The part has been checked, fixed if necessary, and repackaged for sale. The vendor should clearly mark refurbished parts as shown in Figure 1-4. So you often end up with a really nice part for pennies on the dollar as long as you're willing to deal with the potential problems of using refurbished parts. However, in some cases, the price differential is smaller. For example, a refurbished GA-Z97-HD3 motherboard might cost you $98.96, while a new version of the same motherboard might cost $109.99 (an $11.03 savings). You need to decide whether the cost savings is worth the risk. Future chapters discuss this issue in greater detail.

Take the time now to create a preliminary budget for your PC. Think about the goals that you created earlier, the amount you have to spend, and then create a list with some preliminary amounts next to each item. These aren't the final

FIGURE 1-4   Only use vendors that clearly mark refurbished parts.

amounts—they simply represent what you might be willing to spend for that part at some point. Future chapters will help you refine your budget, but having a preliminary budget in mind will help reduce the potential for spending too much on one item and not enough on another, which could result in getting a system that doesn't match your goals.

Identifying the Parts You Have Now

Most people already own a computer or have access to a computer that no one really wants. It's possible that you can reduce costs by reusing some existing parts in your new system. Of course, you always have to proceed with caution. What looks like a potential money saver now could end up costing you later. The following sections tell you about reusable parts in general and how you should identify them for use with the system you have budgeted from the previous section.

Defining What Reusable Means

Not every part is reusable, even if it works just fine. For example, if you want the best graphics performance for a new game, then you need a new display adapter (also called a video card or graphics adapter) unless you already own a high-performance display adapter. Trying to get by with an old display adapter just won't cut it if you want to meet your goals.

One of the more commonly reused parts is the hard drive. The size of hard drives has continued to grow over the years and you may find that your existing hard drive meets your needs. You need to consider whether the existing drive meets the speed requirements for your new system and whether the new system is configured to use it. You see how to perform these sorts of checks in later chapters, but for now you can at least consider whether the drive is large enough to meet your needs. If it isn't, then you've already eliminated the hard drive as a potential reuse part.

It's also possible to reuse your case and power supply in some cases. The case is more likely to work than the power supply, but there are situations where both parts will work just fine in the new system. The biggest thing to consider with the case is whether it's in good shape and is large enough to contain the components you want to put inside it. The power supply must be large enough to supply power to all of the components you want to use.

Some parts are probably going to require change. For example, memory is highly dependent on the motherboard, as is the processor. If you want to build a new system, it's likely that these components are part of the problem that you're trying to solve anyway. You should probably count on replacing the motherboard, memory, and processor as a minimum when creating your new system.

> **NOTE**
>
> *Make sure that any specialty devices that you must keep will work in the new system you build. This means learning the device requirements. Ask the vendor about the part before you make a purchase. Do everything you can to ensure the part will fit. For example, if the part fits in a slot, you need to make sure that the new motherboard has a slot of the correct type. You will discover how to perform this sort of check in later chapters, but for now you need to keep the requirement in mind as you read.*

Obtaining Documentation for Existing Parts

You must have documentation for any parts that you plan to keep. Without documentation, it's going to be hard (or perhaps impossible) to make the old part work with the new system. Fortunately, you can usually get the documentation from the vendor or from a download site online. The documentation helps you perform the following tasks:

- Obtain specifications to ensure that the old part will fit in the new system.
- Configure the part so that it works with the new system.
- Troubleshoot potential problems with the part as you perform system testing.

> **NOTE**
>
> *You don't have to have a paper copy of all the documentation, but it does pay to have a paper copy of the configuration information. It may not be possible to access the electronic copy during the building process (or it might simply be inconvenient). A paper copy also sometimes shows details that the electronic copy doesn't show unless you zoom in. Most vendors no longer supply paper documentation, which means that you'll need to print the configuration information out yourself.*

Ensuring the Part Is Actually Usable

Before you take your old system apart, perform any required tests on the parts that you plan to put into the new system. It often happens that a part is actually bad or doesn't have the functionality that it did when it was new. If the vendor supplies diagnostics, make sure you run the diagnostics to ensure all of the part functions still work.

Chapter 1: Defining What You Want

FIGURE 1-5 Pack any parts you plan to reuse in antistatic bags to keep them reusable.

Most parts will also require software of some type to interface with the operating system on the new system. You may find it hard to locate device drivers and other essential pieces of software for the part. Without the required software, the part may work, but the operating system will never recognize it and you won't ever be able to use it to perform useful work.

Exercise extreme care when removing the old part from the system. Make sure you pack it in an appropriately sized antistatic bag like the one shown in Figure 1-5 until you can move the part to your new system. Otherwise, even a small bit of static can render your useful part unusable by frying the components it contains.

2
Introducing the Major Parts

Chapter 1 discusses a lot of topics that deal with making your custom-built computer a good buying decision. In addition, it helps you focus in on why you want a custom-built system in the first place. However, you might also be a bit confused at this point about some terms in the first chapter, such as motherboard. Even if you have heard of some of the terms in Chapter 1, you may have never seen any examples of what these items look like. This chapter helps correct that oversight. In this case, you get an overview of the major parts (components) in a typical PC. Of course, your PC may contain special components, depending on its purpose, but it will also contain the general components described in this chapter.

> **WARNING**
>
> *If you decide to open your PC case to see how the parts in this chapter compare to those in your system, make sure you don't tamper with any of them. The purpose of this chapter is to acquaint you with the major parts. Future chapters will discuss how to work with the components in a safe manner. Tampering with the components could damage them and make them unusable if you need them in later chapters.*

As you read through this chapter, remember that every part you read about is an opportunity to customize your system. Even the power supply can be customized. Some power supplies provide extra cooling, some are extra quiet, some support special display adapters, and some keep electrical interference under control. In fact, a power supply really isn't the generic part that many people make it out to be. As a result, it's important to think through the descriptions in this chapter to determine whether the part you're reading about might be something you want to customize in your own PC.

Opening the Case

A case holds all of the parts that form the computer except the monitor, mouse, keyboard, and any special external devices you want to attach. The case provides a place to connect everything together and to ensure that all of the parts receive proper cooling. A properly designed case will also reduce electrical interference to some extent and provide you with places to attach future additions, such as hard drives, to your computer. In short, a case acts as a framework for the main part of your PC—the part that this book focuses on most.

In the past, cases were relatively hard to open because they were designed to keep people out and also provide a substantial framework for the parts inside. As parts have gotten smaller and administrators have complained about the time required to open cases, the cases themselves have changed. You can often open a case now without using a screwdriver. If you do need to use a screwdriver, most cases open using a #2 Phillips.

Because of the way in which the parts are placed inside of the case, you normally open the sides. This usually means undoing a fastener of some sort on either the front or the back. The sides can simply slip off as a single piece as shown in Figure 2-1 or you may find that the case uses actual side doors as shown in Figure 2-2.

No matter how you open the case, you see a basic layout of components inside. Figure 2-3 shows a typical layout. The motherboard is against the right

FIGURE 2-1 Some cases open by removing a single piece covering both sides and the top.

Chapter 2: Introducing the Major Parts

FIGURE 2-2 Many newer cases open by removing individual doors—one on each side.

FIGURE 2-3 Most PCs use a similar layout.

side of the case (looking from the front), the power supply is at the top back of the case, permanent storage is in the top front of the case, and the add-on cards appear in the bottom back of the case. You also see one or more cooling fans, one of which is usually at the bottom front of the case. The case layout will become more familiar as you begin to put the parts inside.

Understanding Case Specifications

You may see cases referred to by motherboard type, orientation, size, and material. Most people who build their own systems use the Advanced Technology eXtended (ATX) motherboard, so you need a case that will support an ATX motherboard. If you decide to work with a different kind of a motherboard, then you need a case that supports that kind of motherboard. The "Understanding the Role of the Motherboard" section of the chapter describes motherboard types in more detail.

Cases come in two orientations. The desktop case is longer horizontally than it is vertically and is meant to sit on your desk. The tower case is taller than it is wide and is meant to sit on the floor. Most people today use the tower case, but you can still find desktop cases when needed. The orientation you choose depends on how you want to arrange your computer setup.

Tower cases come in a number of sizes. A full tower is the largest and is about 30 inches tall. It can accommodate between 6 and 10 drive bays where you put hard drive and other types of permanent storage. A mid tower is 18 to 24 inches tall and can accommodate between two and four internal drive bays, along with two to four external drive bays. A mini tower is 12 to 18 inches tall and provides space for two external drive bays and possibly up to three internal drive bays. There are other case sizes that you can obtain, such as the small form factor (SFF) case, but they aren't discussed in this book because they're used for specialized tasks.

The material used to construct your case is also important. The least expensive cases rely on plastic. The problem with these cases is that they don't tolerate much abuse and they leak radio frequency (RF) emissions. What this means is that the new problem with your phone and radio might be your PC. Better quality cases use aluminum construction. They hold up better than plastic cases, are relatively light, and keep most RF inside. The best cases are made of steel. They do tend to be heavy, but they also block all of the RF and last for a long time.

Understanding the Role of the Motherboard

The motherboard is the most important component inside the case. It ties everything else together. You find the memory, ports, and processor on the motherboard, along with a number of other features. Figure 2-4 shows a typical ATX motherboard, which is the type commonly used when building your own PC.

Chapter 2: Introducing the Major Parts

FIGURE 2-4 A typical ATX motherboard.

Although the layout of a motherboard can differ from the one shown in Figure 2-4, the case determines the position of some items. For example, the expansion slots for an ATX motherboard are always located in the back left corner. Any standard ports that the motherboard provides are located in the back right corner. The power supply coupling is normally found on the right side of the motherboard and the connectors used for the front panel switches and indicators are located in the front left. The positions of other items, such as the memory and processor, can vary depending on the vendor design.

Motherboards come in all sorts of shapes and sizes. Many of these designs are for specific needs, such as the mini Information Technology eXtended (ITX) motherboard used with SFF cases for applications where space is at a premium and extendibility isn't required. Avoid using anything but the ATX form factor if you can help it because the ATX form factor has the broadest range of support options. PC builders normally use one of these four motherboard types:

- Full ATX: A 12-inch by 9.6-inch motherboard that is used for most PCs. It typically provides five expansion slots and four memory slots.
- Mini ATX: An 11.2-inch by 8.2-inch motherboard used in smaller PCs. It typically provides four expansion slots and three memory slots. This design

has the advantage of reduced cost and power consumption. However, it usually lacks many of the more advanced features found on a full ATX motherboard, such as temperature monitoring or places for more than one processor. You may have to use this motherboard in your mid tower case because it may be too small to accommodate a full ATX motherboard.

- Micro ATX: A 9.6-inch by 9.6-inch motherboard used in smaller PCs. It typically provides three expansion slots and three memory slots. The smaller design means that you will save even more on cost and on power usage. You may have to use this motherboard in your mini tower case because it may be too small to accommodate a full or mini ATX motherboard.
- Flex ATX: A 9-inch by 7.5-inch motherboard used in really small cases. It typically provides three expansion slots and two memory slots. This is the kind of motherboard you might use in an SFF case designed to accept ATX motherboards for an alarm system or other automation need. Because they're a little rare, this motherboard might actually cost more than a similar micro ATX motherboard and it provides fewer features. However, the small size means you can use it where a micro ATX motherboard would be too large.

> **NOTE**
>
> *If you have a full tower case, it will accept any of the ATX motherboards described in this section. Smaller cases may be limited to smaller motherboards. Make sure you check the case specifications to determine what size motherboard it will accept.*

Some full ATX motherboards do offer multiprocessor support, but they're rare and expensive. If you can't get enough processing power using a single processor with multiple cores (processors within the single-processor chip case), then you might need to resort to a Server System Infrastructure (SSI) form factor, which requires a matching case. You must get a full tower case for any SSI motherboard you want.

Even though the mounting holes and general arrangement of the motherboard is the same as an ATX motherboard, an SSI motherboard requires a larger than normal case to accommodate it, so make sure your case is up to the task before you get one. Some gaming systems and high-end graphics workstations rely on SSI motherboards, which typically come in the following three forms:

- SSI Compact Electronics Bay (CEB): A 12-inch by 10.5-inch motherboard that provides support for two processors. It typically provides five expansion slots and up to eight memory slots. Because of the way in which SSI motherboards are used, you may find that some features, such as built-in networking support, are missing.

- SSI Enterprise Electronics Bay (EEB): A 12-inch by 13-inch motherboard that provides support for two processors. It typically provides six expansion slots and up to eight memory slots.
- SSI Midrange Electronics Bay (MEB): A 16.2-inch by 13-inch motherboard provides support for four processors. It typically provides six expansion slots and up to 16 memory slots.

Providing Power to the System

Some people view the power supply as a generic block that provides power to everything else in the computer. Nothing could be further from the truth. Power supplies vary substantially in capacity and features, just like any other part of your system.

There are a number of features you need to consider as part of any power supply purchasing decision. For example, you must consider the power supply wattage. A power supply must have enough capacity to service the peak load that a system might experience. A system that normally draws 200 W may actually have a peak load of 500 W. In fact, a good rule of thumb is to multiply the running wattage of a system by 2.5 to obtain the peak wattage unless you can obtain peak wattage requirements for each component in the system (which is normally impossible).

The power supply must also provide support for your specific type of motherboard. For example, an ATX motherboard requires that you have an ATX-capable power supply to support it. The reason for this requirement is that the power supply connects to special connectors on the motherboard in order to supply power to it and the expansion boards. If the power supply doesn't have the right setup, the plugs on the power supply won't fit the connectors on the motherboard.

Power supplies are often called upon to provide specialty services as well. The most common service is to provide additional power to the display adapter. Some systems actually contain two (or more) display adapters in a Scalable Link Interface (SLI) configuration. This configuration requires a special motherboard and display adapters. However, you must get a power supply to provide the required power as well. If the power supply doesn't specifically say that it provides SLI support, you shouldn't assume that it does.

One of the most neglected power supply factors is the quality of the power supply. A standard power supply (shown in Figure 2-5) relies on bare wires that can leak a considerable amount of RF energy—causing interference that's really hard to overcome. Higher quality power supplies shield the output cables so that they don't leak RF, as shown with the PC Power and Cooling (lately renamed FirePower Technology) power supply in Figure 2-6. In addition, higher quality power supplies use higher quality components that actually cause the power supply to weigh more, but also ensure that the power supply will continue to provide high-quality power to all of the components in the system.

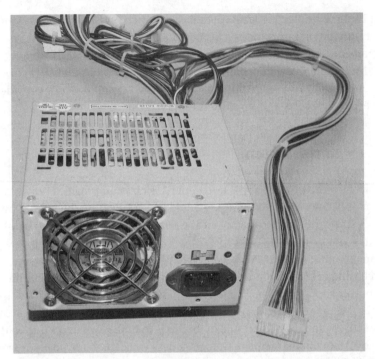

FIGURE 2-5 Standard power supplies use unshielded cables that leak RF energy.

FIGURE 2-6 High-quality power supplies provide cable shielding to keep RF energy under control.

Chapter 2: Introducing the Major Parts

A final consideration for power supplies is the fan they supply. The built-in fan can vary in a number of ways. Some power supplies provide exceptionally quiet fans so that your work environment remains quieter. A power supply fan may also turn off when not needed to conserve power.

Considering the Add-on Boards

Add-on boards (also called expansion boards) enhance the functionality of the basic system. For example, the display adapter is a special kind of add-on board that most systems require. Examples of other add-on boards include network and port cards, both of which are discussed later in the chapter as part of external connectivity.

Better quality motherboards provide a significant amount of functionality. It's entirely possible that you could build a suitable system without using any add-on boards at all. The system would be somewhat basic, but it would provide all the functionality needed to perform some computing tasks, such as setting up a security system. Always take inventory of the features that your motherboard supports before you purchase any add-on boards.

The point is that add-on boards provide some level of system functionality that is either not provided as part of the motherboard or used to complement the functionality the motherboard does provide. The following sections discuss common add-on boards used to modify the way the system behaves (rather than create external connections to it).

> **NOTE**
>
> *This chapter doesn't even come close to describing every sort of add-on board available. There are add-on boards for just about any need you can imagine and probably a few needs you've never even heard of. This chapter covers some of the most common add-on boards. If you need a specialized add-on board, talk with your vendor. You may be surprised to find that the add-on board is not only available, but that there are several configurations of it to meet specific requirements. You also won't find older, generally outdated, add-on boards, such as modems, mentioned in this chapter.*

Working with Display Adapters

Your motherboard may come with a built-in display adapter. This is especially true when buying a motherboard intended for server use. In most cases, the built-in display adapter provides basic display functionality. It most definitely won't fulfill your needs if you plan to use your system for playing games or performing graphics-related tasks such as picture editing or computer-aided design (CAD). However, the built-in display adapter can perform the task if you really are building a server or need a higher-end system to address automation or security requirements. It pays to consider the specifics of the built-in display adapter before you buy an add-on board and turn the built-in functionality off during the configuration process.

Display adapters are rated in a number of different ways. In fact, they're rated in so many different ways that most people find the ratings confusing. The following list provides a simplified view of display adapters that you can use for comparison purposes:

- Output resolution
- Processing speed
- Amount and type of RAM (listed from slowest to fastest type)
  - Double data rate (DDR)
  - DDR2
  - Graphics DDR version 5 (GDDR5)
- Number of output ports
- Type of output ports (with DVI being the most common today)
  - Digital Visual Interface (DVI) (used for high-definition television or HDTV)
  - DB15 (for standard VGA monitor)
  - Video in/video out (VIVO) (used for S-Video television, DVD player, video recorder, and video game connections)
  - High-Definition Multimedia Interface (HDMI) (used for video projectors and televisions)
  - Composite (used for television connections)
- Bus type (with PCIe 3.0 × 16 being the fastest choice)
  - Peripheral Component Interconnect (PCI)
  - PCI Express version 2.0 (PCIe 2.0) × 1 (slowest speed)
  - PCIe 2.0 × 4
  - PCIe 2.0 × 8
  - PCIe 2.0 × 16
  - PCI Express version 3.0 (PCIe 3.0) × 1 (slowest speed)
  - PCIe 3.0 × 4
  - PCIe 3.0 × 8
  - PCIe 3.0 × 16

There are many other statistics you commonly see for display adapters, but this list tells you the statistics that are most commonly needed to make a good buying decision. When buying a display adapter, the output resolution need not be any higher than the resolution of the monitor you plan to buy. The processing speed and amount of RAM combine to tell you how fast the display adapter can process images. Gamers have the greatest need for speed and will

FIGURE 2-7   Display adapters tend to be large, bulky cards.

often spend $500 or more for a single display adapter just to get that last ounce of speed. Most people can get by with a lot less. Anyone who has high-productivity requirements will want more output ports. If you're doing something special such as video editing, then one of the specialized port types, such as VIVO, might be helpful. The bus type of your display adapter must match one of the busses provided on the motherboard or else you won't be able to use the display adapter with the selected motherboard. As shown in Figure 2-7, display adapters tend to be large, bulky cards that contain a lot of electronics (as contrasted to the svelte appearance of sound, host adapter, and other add-on cards).

> **WARNING**
>
> *You can get display adapters for older bus types, such as Accelerated Graphics Port (AGP). New motherboards don't usually provide bus support for these older standards. A low-cost display adapter with the wrong bus type isn't such a good deal after all.*

Display adapters can also come with special functionality. For example, when you perform a lot of video editing tasks or you want your PC to act as the sole input to an entertainment system, you can get a display adapter with a television tuner installed. Chapter 7 describes display adapters in greater detail.

Hearing with Sound Cards

Sound support has become a standard addition for most motherboards today. You can find motherboards without this support, but they're usually not used for standard desktop systems. The built-in sound support usually includes microphone and external device inputs, along with several kinds of output. The support is so good that even gamers rely on the built-in support.

You may find a need for a sound card, however, if you're working with sound at a professional level or if you want to use specialized speaker configurations. For example, some sound cards feature multiple microphone inputs (of the sort you'd need if you're a musician). It's also possible to get cinematic surround sound from an add-in that might not be available with the built-in sound support. Gamers who play certain types of games may find the additional surround sound support useful in detecting the location of an enemy, but you also need the correct speaker setup to make full use of this functionality.

Providing Interfaces Using Host Adapters

Most PCs come with all the interfaces needed to work with modern devices. However, there are situations where you need to install a host adapter to obtain specialized functionality for certain types of devices. For example, installing a Small Computer System Interface (SCSI) host adapter will let you access SCSI devices such as professional quality scanners and external hard drives.

Depending on the motherboard you buy, you may already have some types of host adapters available for use. For example, you can improve the reliability of your system by creating a redundant array of inexpensive disks (RAID). This functionality is also available in the form of an add-in card. Chapter 8 describes various disk options in greater detail.

Controlling Access Using Security Device Cards

Security device cards are a specialized kind of add-on board. You can use them to ensure a user is actually allowed to access the system by reading a fingerprint or other biometric. In many cases, the biometric completely replaces the need for passwords and other easily compromised security measures.

However, security devices can also secure the system or its data. For example, some security device cards encrypt the hard drives in hardware, so it isn't possible to remove the drive from the system and have any hope of using it in another machine. The same technology can encrypt data streams, such as a network connection.

> **NOTE**
>
> *Even though security device cards may seem like an esoteric addition to a computer, the functionality they provide is becoming a lot more common as motherboard additions. The Internet of Things (IoT)—the connectivity of everything to everything else through the Internet—makes security a pressing issue. You may find that you really do need such devices to protect your system in the near future.*

Considering Other Add-on Boards

There are many other special-purpose add-on boards that you can get for a system. One of the most commonly used add-on boards for technicians is the power on startup test (POST) card. As your computer boots, it produces various codes. These codes can be interpreted by a technician with the proper knowledge in an effort to fix system errors that would normally require a lot of trial and error. The technician simply places a special board in your system and reads the POST codes to detect many startup errors.

Some older solid-state drives (SSDs) also come in add-on board form. You simply place the add-on board into your system and it acts as a really fast hard drive. An SSD is typically small, but extremely fast. Some people use them for a temporary drive to speed up disk-intensive tasks or to store commonly used files that the system accesses often. The idea is to make the overall system throughput faster. Newer SSDs look just like standard hard drives, which is why this book covers them as part of the storage device section.

Add-on boards can add any sort of functionality to your PC. For example, many pieces of scientific equipment require a special add-on board (although this need is becoming less common as the scientific community begins to standardize on the Universal Serial Bus or USB connector). The point is that you can probably find the add-on board you need, even if it's of the nonstandard variety.

Deciding on Storage Devices

PCs have access to a considerable number of storage devices today. However, the most common form of storage is still the hard drive because it's relatively fast, inexpensive, and comes in massive sizes. Your system will likely contain more than one type of storage device. Having multiple types of drive available makes your system more flexible and each kind of drive has its own particular characteristics. Table 2-1 contains a listing of the most common storage types and their various features.

Even though it doesn't appear in the table, you can still get the old-fashioned floppy drive for your system. The main purpose for a floppy drive today is to provide a bootable emergency disk for a system. However, you can obtain this same functionality on most new systems using a USB thumb drive. The floppy drive is probably best forgotten at this point unless you have older media applications that require one. There are also other types of devices that don't appear in the list, such as the ZIP drive. Again, these devices were state of the art at one time, but their time is past and it's best to disregard them. Chapter 8 does discuss a number of alternative approaches to using the technologies described in Table 2-1.

TABLE 2-1 Storage Device Characteristics

Storage Device Type	Pros	Cons	Notes
Hard drive	Large storage capacity (size), inexpensive, moderately fast	Not fast enough for some types of applications, prone to damage	The hard drive is likely to remain a favorite due to price.
Optical	Replaceable media, both read-only and read/write formats available	Relatively small storage capacity (size), media damages somewhat easily, media degrades with time, the drives tend to run slowly compared to other storage devices	Optical drives are commonly used for semi-permanent storage, but the shelf life is commonly just 5 years.
Flash drive	Portable, relatively low cost, no special hardware needed	Small sizes make it hard to store much information, extremely prone to damage	Flash drives are a favorite because they work with so many devices and they're incredibly easy to use.
SSD	Moderate sizes, relatively immune to common types of damage, incredibly fast	Extremely expensive, can require special hardware and driver support, only moderately sized devices available, limited number of write cycles	The hard drive will eventually be replaced by the SSD, but not until prices for SSD drives come down considerably.
Hybrid hard drive and SSD	Large sizes, fast, less expensive per MB than SSD alone	Prone to damage, potential for failure due to limited number of SSD write cycles	A hybrid storage device is the easiest and least expensive way to get the benefits of both a hard drive and an SSD today.
Tape	Incredibly large sizes, replaceable media, inexpensive	Incredibly slow reads because most tape drives are optimized for write operations, sequential reads means a huge performance hit	Tape drives appear in this table because they still represent the least expensive method for long-term data archiving.

> **NOTE**
>
> This book doesn't discuss alternative storage strategies, such as storing your data on another system through the Internet. Cloud storage is a software solution that you add to a system after you build and configure it. You also won't find much information about network storage because this solution also relies on access using a fully configured system. However,

it is important to realize that such solutions exist and to take them into account as you define the configuration of your system. For example, using cloud or network storage may mean that you can get by with a smaller local drive.

Connecting with Cables

You need a wire to transfer electricity from one item to another. For example, when you click the light switch in your house, it makes a connection that allows current to flow from the electrical box, through the switch, to the light. The same thing holds true in computers. A wire provides the electrical connection from the power supply to a device. In fact, it usually requires a number of wires to perform tasks. If you had individual wires connecting items in the case, the case would quickly become a nightmarish tangle of wires that no one could sort out. Cables group wires used for like tasks together so that you can make a single connection from a device to its source of power, or between devices to exchange information. Cables may not seem very exciting, but they're an essential part of every computer system.

The reason that cables occupy such an important place in this chapter is that too few people realize how important they are. In order to build a PC, you must have a sufficient number of cables of the right length to make connections between the various devices.

The power supply normally provides enough cables of the right length to provide power to the devices. If you find that you don't have enough power cables, you can obtain a power supply Y (also called a splitter) like the one shown in Figure 2-8. Likewise, if you find that the power supply cable isn't long enough, you can obtain extensions.

> **WARNING**
>
> *Connecting too many devices to a single power supply cable will overload that cable and potentially cause damage to the power supply or the connected devices. Use a Y only when necessary and when you need to use multiple Ys, make sure each is connected to a different power supply output cable.*

Power supply cables are relatively common and straightforward. Data cables are often a different matter. When working with data cables, you must ensure that the data cable has the right ends and that it is produced using the right kind of cable. Even if two cables look precisely the same, their characteristics can vary, so it's important to rely on the cable part numbers to ensure you have the correct cable. For example, Serial Advanced Technology Attachment (SATA) hard drives

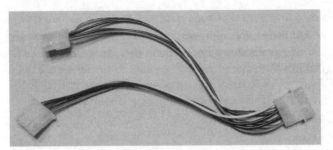

Figure 2-8 Use a power supply Y to connect multiple devices to a single power supply cable.

come with either 3-Gbps or 6-Gbps transfer rates. A 6-Gbps cable will work with either hard drive because its specifications equal or exceed those required to perform the task. However, the 3-Gbps cable could reduce the transfer speed when used with a 6-Gbps drive because the 3-Gbps cable hasn't been designed or tested for the higher speed level. This doesn't mean the 3-Gbps cable always causes reduced data transfer rates, simply that it isn't tested to hold up under the heavier load (or even work with it in the first place). Getting the right cable ensures that your system will behave as expected and that you won't have any surprises somewhere down the road when the less robust cable fails.

> *WARNING*
>
> *It's always important to check specifics when dealing with cables. Less scrupulous vendors will try to sell you a less expensive cable that doesn't meet specifications in order to get the sale based on price. Paying a little more for the right cable almost always saves you time and effort. Consider the slightly higher cost insurance against failure due to conditions beyond your control.*

Keeping Things Cool

For the most part, computers rely on a combination of fans and cooling fins to keep things cool. Devices have cooling fins mounted on the hottest parts. These cooling fins provide greater surface area to dissipate heat. The heat is removed from the cooling fins by air circulated by fans. The larger the cooling fins, the more heat that is dissipated from a given device. Likewise, more fans usually spell greater heat removal.

> *WARNING*
>
> *One of the most common mistakes that people make when dealing with cooling issues is to leave the cables untidy. Cables that are flopping*

around in the center of the case divert and reduce the airflow needed to keep components cool. In fact, it's possible for a cable to become physically entangled in a fan and cause it to fail (sometimes causing the cable to fail as well). Always ensure you keep cables neat and out of the center of the case to maximize airflow. Otherwise, it doesn't matter how many fans you have, some components will overheat.

Some processors also rely on liquid cooling systems, sort of like the radiator used in your car. These processors and their cooling systems are quite expensive. However, they do provide that last bit of processing power that gamers love. Chapter 5 discusses liquid cooling in greater detail. With this exception, you won't find much more than cooling fins and fans in most computers.

The active cooling component is the fan. Most computer fans are 80-mm case fans. However, you can get case fans in a 120-mm size as well. The 120-mm fan requires a place designed for the larger size, so you can't put a 120-mm fan in an 80-mm fan slot. Your processor and display adapter may use specialized, smaller, fans for localized cooling (in fact, you install a processor fan in Chapter 5). The processor fan is usually replaceable, the display adapter fan isn't. Figure 2-9 shows a typical 80-mm case fan, while Figure 2-10 shows a typical 120-mm case fan.

A fan can have one of two ends on the cable. The first is a standard power supply connector and this is the most common type. The second is a special connector that hooks into the motherboard. This second type is important because it lets you monitor the activity of essential case fans. When the fan fails,

FIGURE 2-9   A typical 80-mm case fan.

FIGURE 2-10   A typical 120-mm case fan.

monitoring software can tell you to replace it, saving the valuable component from overheating.

> **TIP**
>
> *It pays to keep one or two spare fans on hand. Replacing a fan before any system damage can occur will save you considerable time, money, and frustration later. Fans are relatively cheap data storage and other system components can become quite expensive, especially when you lose data as a result of their failure.*

The construction of a fan affects how quiet it is and how much air it moves. Fans are rated in cubic feet per minute (cfm). In general, the higher the cfm, the noisier the fan becomes because the aggressive air movement causes flow noise (the noise of air flowing over the fan blades).

Fans also typically use two kinds of bearings. Sleeve bearings are usually quieter than ball bearings. However, ball bearings are more reliable and fail more gracefully than sleeve bearings do. Most computing professionals rely on fans with ball bearings despite the slightly greater noise they produce. However, if sound levels are especially critical, getting fans with sleeve bearings will produce the desired results. Make sure you keep spares on hand when you select sleeve bearings because you'll get little or no warning that the fan is going bad.

> **NOTE**
>
> *Increasing the cfm of your computer does improve cooling, but also draws in more dirt. If you install additional fans, you must also clean your computer more often. Chapter 16 discusses cleaning techniques and helps you understand the importance of keeping your system free of dust (in as much as possible).*

Focusing on External Connectivity

At one time, many (if not most) PCs were stand-alone systems. They really didn't connect anywhere—at least, not all the time. Today, PCs are connected all the time, even when no one is using them. There are two main kinds of connection you need to consider when designing your PC. The first is the connection to the outside world, which normally takes place through a network card. The second is the connection to external devices, which takes place through port cards. The following sections describe both connection types.

Dealing with Network Cards

Most decent motherboards come with at least one network card today. In fact, you can easily find motherboards with two network cards installed so that you can create separate connections to the Internet and a local area network (LAN). Motherboards with even more network capable features aren't unheard of, but if you can find a good motherboard with two network adapters, you should have everything needed for just about any imaginable scenario.

It's unlikely that you'll encounter anything other than an Ethernet connection today. At one time, computers did support a variety of other network types, but Ethernet has won the day. The only time you might actually need to buy a network card is if your LAN relies on something other than Ethernet connectivity (incredibly rare, but it could happen).

Ethernet connections are generally rated by speed. Most home systems today rely on 10/100-Mbps connections, which means the connection can run at either 10 Mbps or 100 Mbps. However, it's possible to find motherboards that have 1000-Mbps (or gigabit) connectivity. In order to actually use this functionality, you must have the right kind of network cabling and other devices (such as routers) that are capable of handling the higher speed. A full discussion of Ethernet connectivity is outside the scope of this book, but you can find a useful tutorial at http://www.lantronix.com/resources/net-tutor-etntba.html.

Dealing with Port Cards

PCs come with a number of ports. A port is simply a connector on the motherboard or an add-on board that provides connectivity for something like a camera, scanner, or printer. In fact, given the right kind of port, you can connect all sorts of things. Your motherboard will generally provide all the ports you need. However, you may find it necessary to buy an additional port card if your motherboard doesn't provide enough of the right types of ports. Here is a listing of the kinds of ports that you typically see on a PC:

- Keyboard/mouse: A round connector used to connect a keyboard or mouse to your system. The connector is usually marked to show which device it accepts.
- Serial: Typically a nine-pin male connector used to attach an older mouse, some types of ancient modems, or a printer. Some scientific devices also rely on a serial connection.
- Parallel: A 25-pin female connector used to attach high-speed devices—normally printers.
- Universal Serial Bus (USB): A modern connector used to attach just about anything to your system including camera, scanner, mouse, keyboard, or printer, anything else you can think of. USB ports come in several versions (all of which use the same physical connector):
  - 1.x: The oldest version found only in older machines.
  - 2.0: A common version found in many machines today.
  - 3.0: The type that your new motherboard will probably have installed.
  - 3.1: A high-speed version that you might find on better quality motherboards.

> **NOTE**
>
> *USB negotiates version number characteristics and speed automatically when you plug a device into it—having a motherboard with a higher version number port setup means that you can use newer devices at full speed. However, a newer port won't make your older device run any faster—the port simply reconfigures itself to accommodate the older device. In some cases, a new device that is expecting a higher version number port won't run in an older port. For example, a USB 2.0 device may not work when plugged into a USB 1.1 port.*

3
Considering the Vendors

After you get an idea of what you want to put into that case, you need to start thinking about which vendor you want to use to obtain the various parts. However, there is a problem. Every vendor thinks it has the best product out there (and may stretch the truth to make it so), reviewers often have biases that affect their reviews, product data sheets have inconsistent and conflicting information, and it's not possible for you to hands-on test every part. Getting the parts can be the hardest part of your building experience because there is always that nagging feeling in the back of your mind that you're not getting a very good deal, that you're leaving things out that your budget won't allow you to get later, or that you'll regret a decision because you weren't quite aggressive enough. If it helps, every other PC builder goes through the same experience. This chapter is designed to help remedy the problem—if not completely, at least enough so that you can sleep at night.

The main issue is to simplify the task as much as possible. For example, rather than try to get all of the statistics to compare, just compare those that are essential to your particular needs. For example, you might not need to know how fast a display adapter can draw sprites if you aren't playing games. However, everyone needs to know that a display adapter has a particular resolution. Comparing just the statistics you need clears the air a bit so that you can better understand which display adapter will suit your needs. In fact, you can often simplify things enough to find several parts that will fulfill your needs—making the decision the simple one of which part costs less.

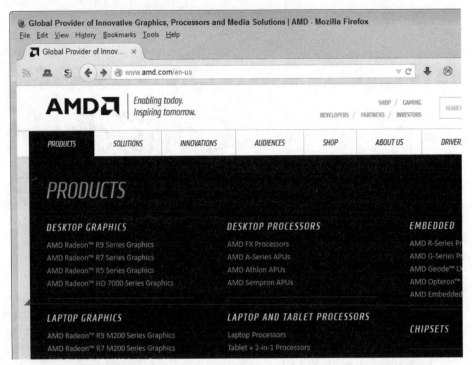

FIGURE 3-1 Vendor sites provide you with a complete list of their product offerings.

Getting on Vendor Sites

The best place to get information about your computer parts is on the vendor sites. The problem with some third-party sites is that they sometimes have inaccurate, missing, or assumed statistics. In addition, when you go to a third-party site, you usually don't see all of the products that a vendor has to offer. For example, when going to the AMD site, you can see that it offers a number of processor types as shown in Figure 3-1.

Making Sense of What You See

Look at the Desktop Processors list and you will see a number of entries that end in APU. Vendor sites will often contain unexplained acronyms—it simply assumes that you know what they mean, which may not be the case. Take the time to look up terms that you don't know before you start poking around too much.

An accelerated processing unit (APU) contains both the central processing unit (CPU) and the graphics processing unit (GPU), so you don't need to buy a display adapter when using an APU chip. However, in order to get that convenience, you give up quite a lot in the way of performance. An AMD APU contains a maximum of four internal processors, called cores, while a CPU, such as the AMD FX series, contains eight cores—twice the computing potential. In addition, the best Radeon GPU you get is an R7. If you buy a separate GPU, you can get the more powerful Radeon R9.

When using an APU, both the CPU and the GPU share a combined memory store. Again, this means that you sacrifice performance to obtain the single-chip solution. Relying on graphics-specific memory means you get a performance boost and there is no contention for the memory store.

As you can see, AMD offers both desktop graphics and desktop processors (among other things like memory). The processors are grouped by family. In order to see a specific processor, you must select the family entry and then choose specific processors to view. The following sections help you understand how to use the information on vendor sites.

Performing Quick Comparisons

Most vendors will provide you with a table where you can perform a quick comparison. For example, if you decide to use an AMD FX processor in your computer, then you can view the Model Number Comparison shown in Figure 3-2. The best processor normally appears at the top. Every FX processor

| Liquid Cooling System | Product Brief | Features | Processor-in-a-Box | **Model Number Comparison** |

AMD FX PROCESSORS

Model Number	Frequency	Total L2 Cache	L3 Cache	Packaging	Thermal Design Power	CMOS Technology
FX 9590	4.7/5.0 GHz	8MB	8MB	socket AM3+	220W	32nm SOI
FX 9370	4.4/4.7 GHz	8MB	8MB	socket AM3+	220W	32nm SOI
FX 8370	4.0/4.3 GHz	8MB	8MB	socket AM3+	125W	32nm SOI
FX 8370E	3.3/4.3 GHz	8MB	8MB	socket AM3+	95W	32nm SOI
FX 8350	4.0/4.2 GHz	8MB	8MB	socket AM3+	125W	32nm SOI
FX 8320	3.5/4.0 GHz	8MB	8MB	socket AM3+	125W	32nm SOI

FIGURE 3-2 Processor tables usually list the best processors first.

contains eight cores, so the table doesn't contain this redundant information—you need to read the product description to obtain it. Notice the Packaging column. This column is important when choosing a motherboard because the packaging determines what sort of motherboard you must get. A socket AM3+ processor will only fit on a motherboard with a corresponding socket.

Obtaining Part Information in the Correct Order

It's possible to waste a lot of time figuring out what you want when you don't follow a specific order in visiting the vendor sites. The following list provides you with a suggested order for making your selections:

1. Choose a processor. The processor you get determines the basis for just about everything else in the case. The two main choices are Intel and AMD. In general, Intel processors offer higher processing speeds at a premium price. AMD is usually the choice for price-conscious buyers.
2. Choose a case. Actually, you could choose the case before the processor, but the case must be able to fit wherever you want to store your computer. The "Understanding Case Specifications" sidebar in Chapter 2 will help you make a good choice.
3. Choose a motherboard. The motherboard processor socket must match the processor you choose and the motherboard size must match the case size. There are many motherboard vendors, but the top four motherboard vendors are: ASUS, Gigabyte Technology, Intel, and MSI.
4. Determine how much RAM you want and the type of RAM you need. A Linux system can get by with 4 GB of RAM without too much problem. When working with a Mac or Windows system, you want at least 8 GB of RAM. Gaming systems normally have 16 GB of RAM. The type of RAM you must purchase is determined by the motherboard you select.
5. Determine which display adapter you need depending on the kind of processor you get and the type of work you want the system to perform. If you purchase an APU, then you likely won't need to buy a display adapter. (Most APU setups provide a configuration option for disabling the APU so that you can add a display adapter later.)
6. Choose any additional add-on cards you require. The kinds of add-on cards you need are determined by the feature set of your motherboard and the kinds of work you perform with the system. Chapter 2 describes various types of add-on cards you can get.
7. Select the storage options for your system. Most systems today contain at least one hard drive and at least one optical drive. Chapter 2 describes

Chapter 3: Considering the Vendors

some of the storage media options at your disposal. For example, you may decide to get a tape backup if your data is exceptionally valuable.

8. Determine how many fans to obtain for your system. The power supply always provides one fan, but you need additional fans in most cases. Here are some of the considerations for choosing fans:
   - Every system should have one fan in the front to help draw in cool air.
   - Add a fan if you have a full tower case because the larger case will likely suffer from poor airflow.
   - Add a fan for every storage device beyond the first two.
   - Add a fan for every display adapter beyond the first one.
   - Add a fan for every physical processor beyond the first one (multiple cores don't count as additional physical processors). (Generally speaking, the system will come with a fan for the first physical processor, but you want to check to ensure that this is the case—every processor does require a heat sink and fan.)
   - Add a backup fan for critical systems.

9. Calculate the amount of power your system will use if possible (including any external devices that you plan to attach to the system that will derive their power from the system through a port or other means, such as a USB hard drive). Multiply this value by 2 to obtain the wattage of the power supply you need. Here are some rules of thumb you can use when determining the size of power supply:
   - Low-end systems with a single four-core processor, two storage devices, a low-end display adapter, up to 8-GB RAM, and just one fan in addition to the one in the power supply can usually get by with a 400-W to 500-W power supply.
   - Medium-power business systems with a single eight-core processor, two storage devices, a moderately powerful display adapter, up to 8-GB RAM, and two fans in addition to the one in the power supply need a 500-W to 700-W power supply.
   - Medium-power workstation with a single eight-core processor, four storage devices, a high-end display adapter, up to 16-GB RAM, and three fans in addition to the one in the power supply need 650-W to 800-W power supply.
   - Gaming system with a single eight-core processor, three storage devices, two high-end display adapters in an SLI configuration, up to 16-GB RAM, and four fans in addition to the one in the power supply need a 750-W to 1000-W power supply.
   - High-power workstation with a dual eight-core processor, five storage devices, two high-end display adapters in an SLI configuration, up to 24-GB RAM, and five fans in addition to the one in the power supply need a 800-W to 1200-W power supply.

> **TIP**
>
> *If you want to perform a more accurate power supply calculation, use the calculator found at http://extreme.outervision.com/psucalculatorlite.jsp. In order to use this calculator, you must have specifics about your system, including items like the specific processor type, number and type of storage devices, and any add-on boards you need to install. The form looks complicated, but if you work through it one item at a time, it's possible to come up with a fairly accurate power requirements value. Get the next largest power supply to ensure you have enough power for the system. Some vendors, such as FirePower Technology, also provide guidelines on which power supplies to purchase based on the load placed on the system (see the Shopping Tools links on the page at http://www.firepower-technology.com/power-products/).*

Performing Apples-to-Apples Comparisons

As you build your list of parts, it's essential to keep the statistics in mind. You can be sure that vendors will bend statistics to their will. For example, there was a long running battle between AMD and Intel. The AMD processor design allowed it to perform more work at a given processing speed than an Intel process. As a consequence, an AMD processor running at 3.0 GHz might perform as well as an Intel processor running at 3.5 GHz. In order to make the public aware of the disparity, the AMD processor model number actually included the Intel equivalent clock rating, such as the AMD FX 3500. The issue caused so much confusion that a lot of people wrote articles about it, such as this one: http://www.geek.com/chips/why-amd-mhz-dont-equal-intel-mhz-557914/. In short, the clock rate statistic of a chip doesn't tell you as much as you might think it will. You must look at the chip as a whole and consider issues such as the number of cores and so on. The following sections provide you with guidelines on getting the most out of your comparisons.

Understanding the Value of Testing

The only reliable way to compare the sheer computing power of chips is to run a test. Of course, most of us can't afford to run tests on a myriad of processors, much less come up with the time to perform the test. Fortunately, others have already done the work. You can see the output of such tests at https://www.cpubenchmark.net/. For example, Figure 3-3 shows the output of tests run on high-end processors. As you can see, Intel chips generally come out on top when it comes to sheer processing power. Intel's advanced design also produces smaller, more power efficient chips. However, AMD continues to be the winner when it comes to cost.

Chapter 3: Considering the Vendors

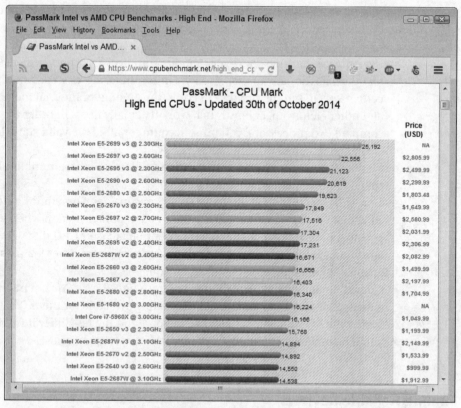

FIGURE 3-3 The only way to actually compare processing speed is to run a test.

Issues surrounding processors are especially noticeable because it's so easy to skew the statistics in so many ways. However, the same comparison issues do surround other sorts of parts. The display adapter is another part that often finds people debating the relative merits of seemingly miniscule differences. Where there are two or more vendors making a part, you find each vendor stating that its part is best. However, it's important to realize that most people will never notice these differences in a built machine. The problem is that applications aren't built to maximize the potential of any given part, and even if they were, the human user often can't make optimal use of the full machine potential.

Performing Comparisons Efficiently

To keep your sanity you need to simplify the comparison process and ensure that you're performing a valid apples-to-apples comparison. The following list provides guidelines for performing comparisons in a way that will help you see actual differences and understand when the differences might not make much difference to you.

- Use only the statistics that every vendor embraces for a part and ensure that the statistics are applied evenly. For example, a processor runs at a certain speed, but the number of cores, the amount of internal memory, and other factors affect how much work it actually does. In order to compare two processors, you must account for all of the valid statistics: clock speed, cores, and internal memory.
- Find testing charts whenever you can to validate your assumptions about speed or other performance issues.
- Don't allow speed to be the only factor you consider. Sometimes a slower APU that saves you a lot of money for a better storage system when you perform a lot of disk-based tasks is better than a faster CPU that will strain your budget too much. In other words, compare other features to ensure you're getting the best deal for you.
- Consider reliability as an important performance factor. Most parts have a mean time between failures (MTBF) rating available for them. The higher the MTBF, the more reliable the part and the less likely it is that you'll be replacing it sometime soon.

Finding Reliable Reviews

A review is simply an opinion in disguise. Opinions need not be based on fact and are often colored by personal biases. For example, a tester who really likes AMD and despises Intel may find faults in an Intel offering without finding those same faults in the AMD offering. The reviewer may not even be aware of the discrepancy—in fact, most aren't, which is why good reviewers need equally good editors to challenge notions of inequality in reviews. Reading a review of a part on Joe's Parts Review may seem like a good idea until you consider the fact that you don't have any idea of who Joe is (he may be an employee of the company that seems to win most of the reviews) and that you don't know what criteria he used for the review.

Good reviews will not only be edited for potential bias (and even then, some bias gets through the editing process), but they also include the criteria used to evaluate parts. If the parts are tested to determine their worth, the tests are made publicly available so they can be examined for bias. In short, a good review should present you with the environment in which the review is made so that you can determine just how much you believe the reviewer. The following sections provide additional ideas on how to get a good review.

Chapter 3: Considering the Vendors

> ***WARNING***
>
> *It's never a good idea to accept a review by someone who represents any of the parties in the review in any manner. A review of an AMD part by a former Intel employee is almost certainly going to be biased in some way. The best reviewers are independent and don't accept any sort of gratuity from the company being reviewed.*

Getting Good Professional Reviews

Some reviewers are professional. In other words, all they do is provide reviews of parts. A professional reviewer makes a living by providing reviews of products, so you're likely to get an informed opinion. Even so, you need to question the review before you accept it at face value. Reviewers are human too. With this in mind, here are some places you can look for reliable reviews:

- Gamespot: http://www.gamespot.com/
- PC Gamer: http://www.pcgamer.com/
- PC Magazine: http://www.pcmag.com/
- PC World: http://www.pcworld.com/
- TechSpot: http://www.techspot.com/
- Tom's Hardware: http://www.tomshardware.com/

There are other places you can look, but this list should give you a good starting point. All of these locations have been providing reviews for an extended period and are known for their relatively unbiased reviews. They all have established and publicly available testing criteria as well and make the review process known every time it changes. In short, even if the review is biased in some way, you know how it's biased and can compensate in how you look at the review.

Understanding the Value of Points

When working with reviews, you need to question every part of the review. Figure 3-4 shows a review with a numeric value assigned to the part. This number is supposed to make it easier for you to make a buying decision by assigning a value to the part. This part rated a 70 out of a potential 100 points. The first thing you need to know is that it's highly unlikely that any part will receive a perfect 100. The second thing you need to know is that the number is highly subjective and possibly the most biased part of the review. Ignore the number, read the review, and examine the test results to make your own decision.

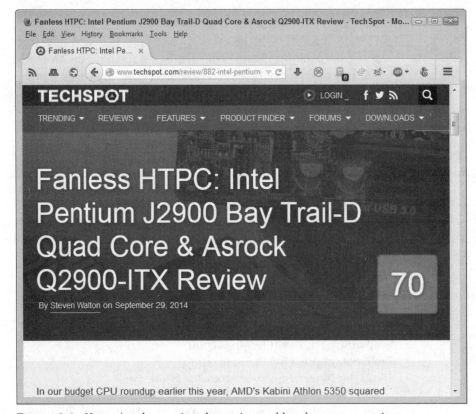

FIGURE 3-4 Numeric values assigned to reviews seldom have any meaning.

Using the Good Review Checklist

The best reviews available usually have certain things in common. Yes, the author's voice will differ between reviews and most authors let their biases show. However, this checklist of items to look for will help you understand when you're getting a quality review that you can actually trust to help you look for parts for your system:

- Test criteria and results are easily understood.
- Testing environment criteria are made public.
- Any tools used in the testing process are listed and the statistics for those tools made known publicly.
- The reviewer commentary makes the reason for specific comments plain.
- The reviewer lists any potential biases as part of the review.
- All reviews for like items follow the same format so that it's possible to compare products.
- The review includes a picture of the product so you can easily identify it.

- There is a list of the product statistics and these statistics are in the same form as other products of the same type.
- A summary includes links to any vendor-specific information about the part.

Understanding the Compatibility Pitfalls

No matter how much material you read and comparisons you perform, there is always a chance that one part won't talk with another. When two parts won't talk to each other, it's called a compatibility issue. Everyone tries to avoid them, but to some extent you really can't as long as vendors try to differentiate products by adding nonstandard features. The problem is made worse when vendors not only add these features, but also fail to fully implement standards or implement them in a way that is guaranteed to cause problems for competitive reasons. As part of your research into parts for your system, you need to consider the potential for compatibility issues. The newer a part is, the less it has been tested, and the greater the potential for compatibility issues to crop up. The following sections help you understand compatibility issues in greater detail.

Considering Problems with Standards Adherence

The price you pay for getting the latest technology is that the standards for the technology haven't been ratified yet (assuming there is a pending standard at all). What this means to you as a buyer is that your part may not meet standards later, which means headaches getting device drivers so the operating system knows how to interact with the part and applications that can use the part. More importantly, it means that upgrades become problematic because new parts will likely adhere to the standards and conflict with the existing part that doesn't. That's the reason those standards listed on the side of the package are so important. You actually need to read them and ensure that the part you're getting follows known standards to ensure a consistent experience and a reduced potential for conflicts.

Unfortunately, even when a vendor says it follows the standards, it doesn't always mean that the vendor has followed the standard completely or that there won't be conflicts. This is where reviews can be helpful. If someone has tested the part in a number of conditions and you can ensure those tests did check for standards adherence, you can get some idea of whether the part will cause conflicts later. Partial standards adherence can occur for a number of reasons:

- The vendor provides noncompliant extensions to the standard in order to gain a competitive advantage.
- The standard is poorly written, so vendors have to come up with an interpretation that may conflict with other vendor interpretations.

- The standard was modified or updated after the vendor already committed to a design for the part.
- A standard may allow for levels of compliance and the vendor chooses a lower level in order to avoid the costs of implementing the full standard.

Considering Problems with Feature Extensions

A feature extension is something that is added to a standard feature. For example, you might find that a vendor adds vibrating functionality to a joystick in order to provide a more realistic experience. Often, this feature will require specialized drivers and applications to use it. In fact, display adapters are especially prone to this kind of problematic addition and will come with special applications so you can see the feature extension at work—mostly because no one else is using the feature extension.

Feature extensions often dazzle the eyes, but bring woe later. As new technologies become available, marketers want to implement them immediately in order to make sales. Products that have a lot of pizzazz attract buyers—whether the technology is ready for general use or not. You can usually spot feature extensions quite easily because the vendor site will discuss them continually.

Having a system with the latest features isn't bad, but you need to exercise care in what you add and realize that you'll have to live with the decision. In many cases, the feature extension won't garner enough attention to grab developer time and it will languish because no one is using it. The feature looked great, but now it's a white elephant that only causes conflict problems with other hardware.

Working Through the Odd Bad Part

People think that computer parts work much like light bulbs—they either work or they don't. Unfortunately, computer parts are much more complex than refrigerators, televisions, or toasters. A computer part can fail in extremely odd ways. One of the most common problems is a solder joint that opens when the joint is cold, but closes when it heats up. So, the part fails to work when you first start the computer, but then it starts to work after the computer warms up. Eventually, the part will fail, but for now, it works when the computer has a chance to warm up.

A single gate can fail within a chip, so that the chip mostly works, but some little function somewhere doesn't work as expected. You may not even notice a problem until you run just the right software that makes use of that gate. Suddenly, your wonderfully stable computer has all sorts of frustrating problems, but only when you use that specific piece of software.

The ground used to keep noise at bay can become less reliable. Perhaps the connection isn't just right, so there is resistance between the part and its ground. As noise levels change in the computer, the part alternately works and fails.

Chapter 3: Considering the Vendors

It may not even fail completely—it may simply work slower or produce unexpected results.

The point of all this is that you need to ensure you understand the return policy for parts, test the system thoroughly after you put it together, and understand that your computer really isn't the same as a light bulb. In addition, hold on to all your receipts for the full warranty period because no one will accept the part back without them.

Dealing with Compatibility Issues

At some point, you're likely to encounter compatibility issues. However, they need not be a source of frustration as long as you understand that there is a reason for the problem and that it's fixable if you remain patient. Here are some techniques you can use to deal with compatibility issues once they do crop up:

- Verify the devices in your system are designed to work together at the outset. If your motherboard only supports SATA II drives, getting a SATA III drive and expecting it to work at full speed will only result in frustration.
- Ensure you have the latest drivers and upgrades. The part you get in the box is often outdated by the time you get it because it has sat around in a warehouse. Getting the latest drivers and updates often cures compatibility problems.
- Check the installation. Sometimes a loose cable or a device that isn't secured properly will cause problems. For example, an add-on board that isn't screwed to the case properly could rock in the slot and cause intermittent errors.
- Contact the vendor about them. Vendors are often aware of the problem already and may have a fix for it.
- Talk with other users. Vendor sites often host areas where users can support each other and vendor representatives often provide helpful information as well.

Reading Between the Lines

Theoretically, a vendor won't lie to you. However, you may find that vendors stretch the truth a bit to get your sale. In addition, there are considerations you need to make in reviewing information the vendor provides:

- Most tests are performed in a lab environment, rather than a real-world environment.
- The vendor will select an environment that is favorable to a particular part's feature set.

- There are differences between each part, so what you read online may not perfectly match what you see in real life even under perfect conditions.
- The performance you see on your buddy's system is unlikely to match the performance on your system (your system may be better or worse, depending on how you use the part).

The bottom line is that you won't actually know how a specific part will perform until you stick it into your machine. You can make a really good guess based on what the vendor specification sheets, reviews, and your buddy's system tell you, but you can't know precisely how the part will work until you install and configure it. The most important concept you can take away from this chapter is that getting what you want from a part often means doing a lot of research and then having a willingness to tweak the part after you get it.

4

Getting What You Need

By the time you reach this chapter, you have spent quite a bit of time thinking about what you want to go into your system. Research is important because a lot of people end up with PCs they don't want because they never put in the time to figure out exactly how they'd use their system and which parts would work best. In addition, each vendor has something unique to offer, so knowing which vendor best suits your needs is essential. Of course, all this research will help you get a better system put together, but it lacks organization. This chapter will help get you organized and also start you along the road toward actually building your PC.

Getting the parts is just part of the process. You also need to obtain the tools required to put the PC together and ensure you have a work area that will actually allow you to complete the work. PCs tend not to require a lot in the way of tools, but there are a few specialized items that come in handy, especially if you have larger hands and find that working in tight spaces is a little on the challenging side. Ensuring you have addressed issues like lighting the work area properly also helps a great deal.

> **NOTE**
>
> *It's important to remember that everyone has different approaches to performing work and that the techniques in this chapter are suggestions on how to get started. You might find that you need a little different setup or that you require a few items that aren't mentioned in the way of tools. The goal is to build a functional PC, which means keeping static under control and not losing those tiny parts somewhere inside the machine. How you accomplish that goal will vary slightly from the*

techniques found in this book, but be sure you keep the goals in mind as you work to ensure you don't end up with a dead PC.

Creating and Verifying a Purchase List

Before you can buy anything, you need a list of items to buy. A lot of people go to the grocery store and wing it, but read just about any article about saving money in the store and you find that one of the tricks is to create a shopping list. You want to stay on budget with your system, which means having that all important shopping list.

However, there is more to creating the list than simply putting the items down. You need specifics about the parts you want to buy. In addition, it pays to comparison shop the parts, which also means figuring out precisely what you're going to get along with the part. Warranty service and return policies vary greatly between stores. It also pays to know the store's reputation for issues such as delivery. With this in mind, the following sections help you create your shopping list.

Dealing with Specifics

One of the problems that you'll encounter is that stores will have a part that's almost the same as the part in another store, but not quite. In fact, without a list of specifics, you can easily end up with a part that isn't anything like the part that you actually wanted. When putting your list together, consider including these sorts of bits of information:

- Part or model number
- Precise part name
- Part vendor
- Amount of memory (or storage capacity)
- Optional components (such as cables)

There are situations where specialized knowledge is helpful. For example, by checking online conversations about a part, you might discover that there are actually several versions of the part floating about and that some versions have issues (such as early failure). It helps to record any specialized knowledge you can obtain about the part so that you don't end up with a part that won't quite do the job.

Considering Reputation

If you search for a part using something like Google, you may end up with a long list of stores that you know nothing about. The store with the lowest price may be reliable and reputable, but it pays to know that they are before you buy from them. A store doesn't suddenly become reputable just because it has an Internet presence. Here are some ways to check the reputation of the stores you want to use to make purchases:

- Length of time in business (stores that have been around longer are usually more reputable)
- Better Business Bureau (BBB) (http://www.bbb.org/) accreditation (having a good reputation with the BBB doesn't come with guarantees, but it does reduce risk)
- Independent positive reviews (those outside the purview of the business itself)
- Published (and verifiable) contact information that includes telephone number and address
- Prompt and courteous responses to e-mail queries

TIP

It generally pays to buy all your parts from one store, which may mean paying a little more at the outset, but experiencing fewer problems later. If you work with a single store, you won't encounter as much of the finger pointing that can occur when something goes wrong. In addition, all your parts will arrive together so you aren't waiting for that one critical part to arrive from a store that is slow to ship. Using a single store can mean getting discounts, such as free shipping or a percentage off your purchase for making a larger purchase.

Performing Comparison Shopping

It's likely that one store will have one or two items on your list at a lower cost than any other store on your list. However, what you really want is to find the single store that has the overall lowest price for all the parts and offers extras, such as free shipping. Most people fail to consider the cost of various extras, such as shipping and handling, as part of the price of buying a part. Some stores will offer you an incredibly low price on a part and then make up for that price by charging you a small fortune in shipping and handling fees.

Also consider the cost of continuing business with a particular store. For example, if you need to return a part, you need to consider the presence of a restocking fee. Some stores will charge you a restocking fee under suspect circumstances (such as when they send you the wrong part).

Verifying the Warranty

Reading the fine print on the warranty is important. For example, the warranty might appear to apply to everything, but the fine print may exclude items that are purchased from a third-party vendor. You have to count on at least one part failure during the building process. Even if you never have a part fail, you must still act as if at least one part will fail. The warranty is an essential part of your contract with the store.

Also ensure that you can return the part to the store. Some warranties require that you absorb the additional expense of working directly with the vendor to obtain warranty service. In other words, once the part leaves the store, you're basically stuck with it even if you have a warranty in hand. A warranty that requires you to send the part back to the vendor often involves a lot of finger pointing and not a lot of satisfaction.

> *WARNING*
>
> *Make sure you read the warranty with extreme care. Some warranties automatically exclude custom-built systems unless the builder is certified by the vendor. In other words, your home-built system won't qualify unless you have the proper certification to work with the part. Vendors often use this technique to ensure experimenters and hobbyists have no recourse in replacing bad parts.*

Checking the Return Policy

The return policy is often filled with hidden issues. For example, the most common problem is with software. You can't return the software once the package is opened, but you can't verify the media is good until you do open the package. Software isn't usually covered by a warranty either. So, you end up with a broken disk and no way to get a useable disk without buying a second copy of the software.

Return policies contain all sorts of monetary issues as well. It's common practice for a store to charge a restocking fee, assuming they accept the part back at all. Make sure you understand precisely what is required to return a part. If a store requires that you send a part back UPS (and by no other route), then you'll need to use UPS in order to ensure its cooperation. You may find that it sounds as if the store will pay postage, but on a second examination of the return policy, you find that the store will provide you with a return postage sticker that they then charge to your credit card when the package arrives.

Making the Purchases

Once you have your purchase list in hand and have verified the various store and vendor policies, it's time to make an actual purchase. The best way to proceed is to purchase things in the order listed in the "Obtaining Part Information in the

Correct Order" section of Chapter 3. The reason you want to order things in a particular order is to ensure that you get the critical parts first. If you can't even get the critical parts ordered properly (perhaps because they're out of stock), then you may want to use a different store to make your purchases.

The following sections help you overcome a few hurdles in your shopping experience and then make the actual purchase. The goal is to ensure that you have a good shopping experience and actually get the parts that you want. Given how much computer parts can cost, you need to ensure you get the right materials from the store and vendors.

Allowing Scripts

Some people are conservative when it comes to allowing scripts to run in a browser with good reason—one virus attack can wipe out more than just your data. Identity theft and other potential problems can become a nightmare. In most cases, it's simply better to disallow scripting, at least on a selective basis. Fortunately, there are add-ons you can use with your browser to keep scripting at bay, remove cookies (including the Flash variety), and block ads. However, when visiting a store site, you'll likely need to allow certain scripts to run. For example, when using Amazon.com to make a purchase, you need to make scripts from these locations active in order to complete the purchase:

- amazon.com (the scripts from the main site)
- images-amazon.com (lets you see what you're buying)
- cloudfront.net (a service that makes it easier to deliver content to your machine)

You normally want to continue blocking scripts from other locations. For example, most sites work just fine when you disable the scripts from doubleclick. net, which is an ad-serving and click-tracking vendor associated with Google. The trick is to make the site usable by enabling the fewest number of scripts needed to get the job done.

If you have a question about a particular site, you can usually get information about it by searching the Internet. However, some ad blockers, such as Ghostery, make the process simple and tell you all about the site as shown in Figure 4-1. In this case, you learn all about Full Circle Studios and its tracking software, ScoreCard Research Beacon. There is never a good reason to allow someone to track you—only cases where you must allow the tracking in order to complete a purchase.

Getting Your Information Together

As you look through sites and figure out what precisely you want to get and how much you need to spend for it, you need to write everything down. Table 4-1

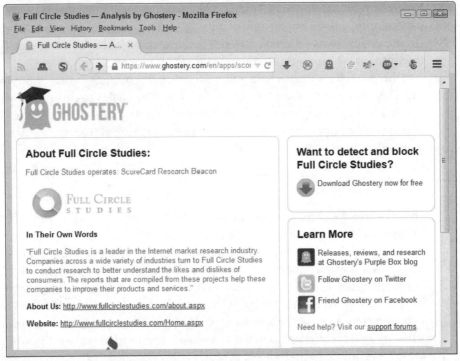

FIGURE 4-1 Ad blockers such as Ghostery can tell you about blocked sites.

TABLE 4-1 Components for the Example System

Component Type	Model	Price
Processor	AMD FD6300WMHkBOX FX-6300 Six-Core 3.5-GHz AM3+ Processor	$100
Case	Sentey Computer Case Cs1-1399 PRO	$40
Motherboard	ASUS M5A97 LE R2.0 AM3+ Motherboard	$79
Memory	Kingston HyperX Beast 2 × 8-GB, DDR3, 1866 MHz	$170
Display Adapter	Sapphire Radeon R9 270X 2-GB GDDR5	$160
Add-on Device	Kinivo BTD-400 Bluetooth 4.0 USB adapter	$14
Add-on Device	Sabrent NT-H802N USB Wireless-N WiFi Network Adapter	$30
Hard Drive	WD Blue 1-TB Desktop Hard Drive	$55
DVD Drive	Samsung SATA 1.5-Gb-s Optical Drive	$15
Power Supply	Sentey Mbp750-hs	$75
Total		**$738**

shows the component list for the computer used for the examples in this book. As you can see, the component list tells precisely what part is needed and how much it costs. The budget for this computer is $750 (part of which went toward shipping costs).

> ***NOTE***
>
> *Table 4-1 doesn't include tax, shipping, and handling. When creating your own information list, you need to take these items into consideration.*

> ***TIP***
>
> *The example machine was purchased from a single store. Adding all the items together made it possible to get shipping and handling free for most of the components (except for the case, which was ordered from a third party). Of course, the site still collected local tax.*

The point is, you need a list with specifics that includes all of the parts as described in the "Obtaining Part Information in the Correct Order" section of Chapter 3. Writing the list down helps you purchase the parts successfully when you actually make the purchase. Even if you've built a lot of systems, you need some sort of checklist because even the experts end up with the wrong part sometimes.

Completing Your Purchase

At some point, you'll start sticking parts in a shopping cart. It doesn't matter whether you're buying parts in an online store or a bricks and mortar store—the essential process is the same. Use your list to buy each part one at a time. Eventually, you end up with a shopping cart full of parts. When you get to this point, make sure you check your cart to ensure your cart matches what you see in your parts list.

> ***WARNING***
>
> *Beware of substitute parts. Just about every site you visit will try to make a substitution if the precise part you want isn't in stock. It's important to check all the details. For example, when buying a display adapter, the model number might match precisely, but the store may try to substitute a part with 2 GB of memory for the one you wanted with 4 GB of memory. Yes, you'll get the part less expensively, but then you also have a part that may not have the efficiency you need to perform tasks. Display adapters are especially problematic because the memory is soldered in*

place most of the time, which means that you can't simply upgrade to the higher level memory by adding more chips. The 2-GB display adapter you bought is the one that you'll keep until you buy an entirely new display adapter.

When you check out, make sure you check the parts one last time. If it sounds paranoid to keep checking, just think of the expression on your face when you open the box and find the wrong part inside. Check, double-check, insanely check a third time—it's all necessary to ensure you have the right products in hand. The checking process also sometimes causes a little light to go on as well—the one that tells you the part won't work because the motherboard or some other component won't support it.

Verifying the Package Contents

If you didn't buy your parts at a brick and mortar store (most people don't), then you'll receive a box at some point that contains part or all of your order. Before you do anything else, you need to inventory the box to ensure you have everything needed to build your PC. The following sections discuss some of the issues you need to consider during the order verification process.

> **NOTE**
>
> *Even if you buy your parts from a single online store, you may end up receiving more than one box. The store may have some parts in stock and send them immediately. Any parts that need to be ordered could arrive later. In addition, some stores, such as Amazon.com, support third parties. Amazon.com won't ship the part to you—you'll receive it from the third party instead. There is also a good chance that the parts won't actually fit in a single box—parts like the case are rather large. In short, don't panic immediately if that first box you receive contains less than the entire list of parts you ordered.*

Checking the Boxes

Take your parts list in hand and check off each item you find in the box. There should be one parts box for each item on your list unless you ordered multiples of some items, such as memory. If your list isn't completely checked off when you're done, check the receipt you should have received from the vendor to determine whether some items are coming separately. If you didn't receive a receipt, check on your account page online to verify the order status. The store will provide some means for you to verify the state of your order. If not (and this would be incredibly unusual), contact the store directly for status information. The big thing is not to panic; the support person at the other end of the line is on your side and wants you to be happy.

> **WARNING**
>
> *It's an extremely bad idea to use a box cutter to open your box. The knife could slice into some of the parts boxes and damage something. Use a shorter knife or scissors to open the box carefully.*

Make sure you keep all of the packaging as you go through the box and try your best not to damage it in any way. You may need the packing material to send any items that are broken or otherwise unsuitable back to the store.

Checking the Individual Parts Box Content

Open each parts box—one at a time. Don't open more than one parts box at any given time because you don't want to mix up the parts and documentation.

There should be a list of what you should expect to find inside the parts box either on the outside of the box or as part of the documentation inside. Verify that each item listed on the content list actually appears inside the box.

> **WARNING**
>
> *Never open any of the actual part packages. The static bag used to hold the part protects it from static discharge. Any discharge could damage the part. Even a discharge that is so small that you can't feel it can (and often will) damage the part. All you want to do is ensure that the part is present and that it's apparently not physically broken.*

When you have verified the content of the parts box, you should carefully repackage it precisely as you received it. You don't want to stress the part in any way. In addition, you want to ensure that all the parts are put back inside the box and that the parts are able to stay inside the box until you need them during the assembly process.

> **WARNING**
>
> *Always wait until you have all of the parts for your computer before you begin assembling it. One of the most common problems that builders encounter is parts that become lost while waiting for a related part to arrive. It's always best to wait until you have everything needed before you start the building process.*

Making a List of Extras

As you go through each parts box, you need to make a list of extras—the items that the vendor forgot to tell you about. For example, the mouse you purchased may need batteries. Yes, they're normally supplied, but not always.

A more common need is cables. For example, a motherboard may ship with only two Serial Advanced Technology Attachment (SATA) cables. The vendor assumes that you want to attach one hard drive and one optical drive to the system. If you plan to attach more devices, then you need more cables.

In some cases, the documentation will actually tell you that a needed part is extra and not supplied with the device. For example, some hard drives don't come with attachment screws. Your case will likely come with an assortment of screws of the right type, but not always. Make certain you have the screws required to attach the hard drive to the case.

A more common problem is adapting 3.5-inch drives to fit in a 5.25-inch drive bay. In this case, you need a special mounting bracket like the one shown in Figure 4-2. (It's possible to find these mounting brackets for all sorts of conversions, so make sure you find the specific type for your need.) You won't actually find this part listed anywhere. The documentation won't tell you that it's required and the need for the part isn't obvious. It's the sort of thing you usually figure out when you try to build the computer. Thinking about how your computer will go together can help you avoid these sorts of problems.

It's becoming a less common problem today, but one cable tends to drive people nuts because it's not even standardized in some cases. The cable that connects the audio output of an optical drive to the sound board

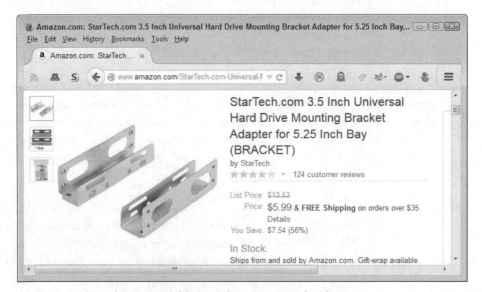

FIGURE 4-2 One of many available hard drive mounting brackets.

Chapter 4: Getting What You Need

FIGURE 4-3 Some audio cables have four different connectors on them to fulfill a variety of needs.

(or motherboard equivalent) on a PC isn't supplied in many cases. The first time you'll figure out that you need it is when you start to play a music CD or video and find out that there isn't any sound available. To keep your sanity intact, you want to get one of the multi-connector cables that fit most needs like the one shown in Figure 4-3.

Getting the Extras You Need

Building a system often requires extras, such as cables, that no one really mentioned at the time you were creating your initial parts list. When you create your extras list, make sure to provide two columns for it: Ordered and Received. Try to buy the parts locally, if you can, to save time.

When you must order online, make sure you check the columns as needed. It's important to keep track of the extras because you can't complete your system without them. However, since the other parts are already on order, you want to order the extras as you become aware of them, which could lead to problems if you're not careful.

Once you receive an extra, it helps to group it with the part that you'll use it with. For example, if you get additional SATA cables, put them with the motherboard or with the drives that will rely on them. You want to be able to find the extra when you need it later.

> **WARNING**
>
> *Don't put the extra in the box with the part. The box is designed to hold the part in a certain way and if you put the extra in the box, it could create stress that will damage the part (especially when it comes to boards).*

Ensuring the Documentation Is Complete

Depending on your situation, you might find that you can't access the Internet as you build your new system, especially if you're using parts out of your old system to do it. No matter what situation you find yourself in, it always pays to have a complete copy of all the documentation for your part and to study that documentation before you start doing anything. Some vendor documentation is poorly written and you need to understand what you're doing before you do it.

Most complex parts will come with more than a quick start guide. This is especially true of motherboards. If all you receive is a quick start guide, then you need to obtain the full documentation for the part on the vendor site. Figure 4-4 shows an example of a vendor site. Notice the Manual & Document link on the

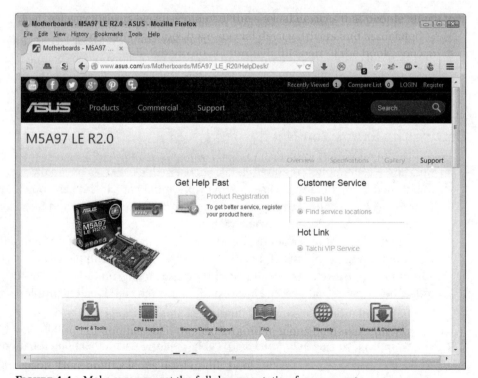

FIGURE 4-4 Make sure you get the full documentation for your part.

right side of the bar at the bottom. This is where you'd click to obtain the documentation.

It helps to have printed copies of some of the documentation. Of course, you probably have a printed copy of the quick start guide already. However, the quick start guide will assume that everything is going to go according to plan and you have no special needs whatsoever. In addition, it assumes that you can actually find that plug you need to use on the board. You don't have to print out the entire guide, but it helps if you have printed copies of these items:

- Overview of the board with its components and callouts to the important items
- Special assembly procedures
- List of added items (such as cables)
- Troubleshooting guide

Inventorying the Required Tools

Your case literature says that you won't need any tools to work with it—don't believe it. When assembling a computer, you really do need tools. Technically, you can probably get by most of the time with a #2 Phillips screwdriver and a flashlight, but other tools can be helpful (and sometimes necessary). The following sections describe two methods of getting the tools you need together.

Obtaining a Computer Toolkit

A computer toolkit is the easiest method of getting the tools you need. In most cases, they include specialized tools. Although you'll seldom need one, a chip puller can come in handy at times. A screw grabber comes in even handier at times, especially when you drop a screw into an area that would require you to disassemble part of the computer. Figure 4-5 shows a typical computer toolkit that you can buy at almost any computer store that sells parts or online at places such as Amazon.com. Table 4-2 describes each of the tools shown in the figure.

> **WARNING**
>
> Some less expensive computer toolkits contain magnetic tools. These tools can damage magnetic media on your system. It's always a better idea to pay the extra for a nonmagnetic toolkit. The tradeoff is that more of the tools will be plastic instead of metal and it'll be easier to lose screws off the tip of the screwdriver.

Part I: Developing a PC Plan

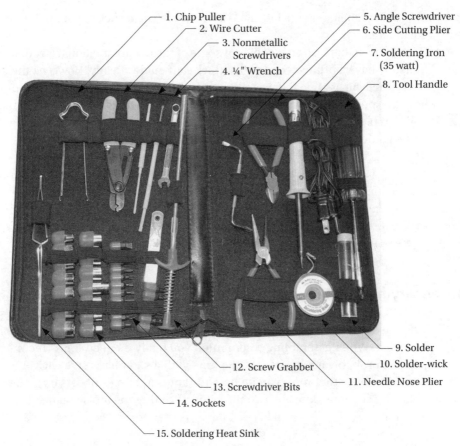

FIGURE 4-5 A computer toolkit often includes specialized tools that you might need in a pinch.

TABLE 4-2 Computer Toolkit Tool Listing

Callout	Tool	Purpose	Essential?
1	Chip puller	Reduces the risk of damage to chips when removing them from a board	No (but highly recommended)
2	Wire cutter	Cuts wire and insulation when attaching new wires to the system components	No
3	Nonmetallic screwdrivers	Makes it possible to adjust tuning screws on analog components without interference	No
4	¼" Wrench	Reduces the risk of damage to circuit boards when removing nuts and bolts	Yes

Chapter 4: Getting What You Need

TABLE 4-2 Computer Toolkit Tool Listing (*Continued*)

Callout	Tool	Purpose	Essential?
5	Angle screwdriver	Allows removal of screws in tight places, such as the top of a case when working with a hard drive	No (but highly recommended)
6	Side-cutting plier	Cuts wires and cable ties	Yes
7	Soldering iron	Allows addition and removal of soldered components from circuit boards, repair of wires to components such as speakers and switches, and potential repair of case elements	No
8	Tool handle	Accepts various tools such as screwdriver bits and the socket holder	Yes
9	Solder	Used to repair electrical connections	No
10	Solder-wick	Removes solder from a component	No
11	Needle-nose plier	Allows grabbing of small objects in tight spaces (such as jumpers)	Yes
12	Screw grabber	Provides the means for accessing small items in the case and picking them up	Yes
13	Screwdriver bits	A set of various bits that are inserted into the Tool Handle (item 15) and used to remove screws of various sorts. The #2 Phillips bit is the most commonly used.	Yes
14	Sockets	A set of sockets used to remove bolts from the case and from cards (the ¼" socket is most commonly used)	Yes
15	Soldering heat sink	Attaches to a component you want to solder in order to protect it from too much heat	No

Getting Individual Tools

You may choose not to get a toolkit. Perhaps you have all the tools you really want for right now. Even though the toolkit is convenient and provides an easy way to carry your tools around, you can build a perfectly acceptable toolkit out of individual tools. Here are the must-have tools that you really should have in any toolkit:

- #2 Phillips screwdriver
- Side cutter plier
- Needle nose plier

- Screw grabber
- ¼" Wrench
- Small socket set

Tools such as the chip puller are convenient, but you won't use them often (or possibly at all). For example, the chip puller can reduce potential damage to the processor, should you ever need to remove it, but generally, you only insert the processor once and never remove it. Gone are the days when computer builders regularly worked with chips, so this tool doesn't see a lot of use today.

Setting Up a Work Area

You can't build a computer while sitting in a cramped, messy area and expect to be successful. I'm sure someone must have built a computer in such conditions, but for the most part, having the right work area is a better idea. The following sections give you some ideas for setting up your own work area. They don't prescribe a precise layout, but act as guidelines on a setup that works well.

Selecting a Worktable

Your worktable should be large enough to accommodate the case and sturdy enough to support it without wiggling. Yes, it's possible that you can get by with two card tables put together to form a single table as long as the card tables are sturdy enough, but it isn't the ideal setup. Some people use a banquet table with folding legs—the heavier type that you often see in restaurants or in catered venues. My own worktable is actually in the kitchen as shown in Figure 4-6. It's the perfect height and quite sturdy.

Notice that there is room for the case, my toolkit, and some of the parts for the example system. Everything is laid out for ease of access. The table width makes it possible to position the case just as it will be laid out when the system is functional. There is absolutely no clutter in this setup to cause problems with lost parts, part damage, or potential sources of dirt.

Ensuring You Have Enough Light

Computer parts are small and easily damaged. If you don't have enough light, you're almost certainly going to bend something, like the pins on your processor. A light right over the top of your work area usually works best. Figure 4-6 shows the light used for my setup. It has multiple bulbs in a single fixture so there are fewer shadows and sits right over the top of my work area.

Chapter 4: Getting What You Need

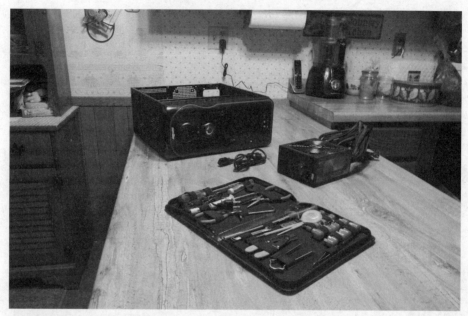

FIGURE 4-6 A kitchen worktable can be the perfect place to put your computer together.

Checking the Outlets

You need at least one easily accessible outlet while building your computer. Even though you won't normally keep your computer plugged in while working on it, you do need to test it from time to time. In addition, you can discharge any static electricity using the power supply case as long as the power supply is plugged in. In my case, I have plenty of outlets, one of which is directly accessible at the end of the worktable as shown in Figure 4-6.

Keeping Your Work Area Clean

Small screws and other components are sometimes dropped during the building process. No, it's never a good idea to drop something, but it does happen. You want to be able to find the screw quickly and easily. Unless your work area is clean, you'll find it difficult or impossible to find the tiny parts you drop. Keeping things clean will reduce accidents, reduce the potential for mistakes, and increase your chances of getting a functional system with the least effort possible.

Part II
Building the Hardware

5

Adding RAM and Processor

You have a shiny new motherboard, RAM, and processor, along with a number of other components. It's possible to install the motherboard into the case first, but easier if you install the RAM and processor before you install the motherboard. Having the motherboard in a place where you can access it easily makes the process faster. With this in mind, this chapter is going to get you started on the building process by adding RAM and the processor to the motherboard.

Understanding Static Electricity

It's a dry day and you've walked for some distance across the floor in your slippers. When you touch the doorknob, you will see an arc race from your finger to the doorknob and feel the jolt of static electricity. Some people view static electricity as a little thing, but it really isn't that little. Think of it as a micro-lightning bolt. Most people don't realize that the static they discharge ranges from 4000 to 35,000 V, which is more than enough electricity to fry some types of electronics, including your motherboard, processor, and RAM.

When you open the boxes containing the components, you'll notice they're carefully wrapped. The plastic used to wrap the electronics has small filaments in it to protect the electronics inside from static discharge. In other words, they're antistatic wrappings. You must use antistatic plastic to transport components anywhere. Otherwise, you will risk damaging the component with static.

Before you can open the packages, you must discharge the static that your body has built up. One way to do this is to buy a special wrist band, attach it to a ground, and then put it around your wrist. The problem is finding a grounding source in most home settings. As an alternative, you can purchase a wireless version of the same device, but these devices require time to discharge the static and it isn't possible to be absolutely certain that the device has done its job (some people report it takes up to 30 minutes before you're static free; at least one review, see http://www.esdjournal.com/techpapr/sfowler/wireless.htm, says they don't work at all). The least expensive and surest way to perform the task is to plug the power supply in and discharge the static in your body by touching the power supply case (and no other part of the power supply). Just one touch is enough. Unplug the power supply when you finish discharging the static.

After you discharge any static, you can work with the components. If you aren't wearing a wristband, walking around could build up a new charge, so you need to discharge any static any time you start working on your system again. Keeping static charges at bay is an essential part of working on components safely. You'll find notices about static electricity throughout the book as reminders, but anytime you touch any components it's a good idea to don a wristband or touch the power supply to get rid of static electricity. The alternative is to lose perfectly good components to electric shock.

Verifying the Processor and RAM Positions

If you have peeked at your processor and RAM, you can see that they're a bit on the delicate side. Figure 5-1 shows the pins on the processor and Figure 5-2 shows the fingers on the RAM. Damaging either one will force you to buy a new item.

FIGURE 5-1 The processor pins are extremely delicate and easy to bend.

Chapter 5: Adding RAM and Processor 71

FIGURE 5-2 Even though the RAM is a little sturdier than the processor, it's still easy to break.

It's nearly impossible to fix either item once damaged. For this reason, you want to ensure you put the processor and RAM in correctly the first time—you really may not get a second chance.

> *WARNING*
>
> *If you bend a pin on the processor or damage the RAM in some way, you can't insert it into the motherboard. Doing so could cause damage to the motherboard, power supply, and add-on boards.*

The best way to work through the building process is to identify the locations of sockets on the motherboard before you do anything and to read the directions that come with the components to ensure there are no special issues you need to consider. Figure 5-3 shows the locations of the processor and RAM sockets on the example motherboard. At this point, make sure you look at your motherboard and use the manual you obtained for it to ensure you know the locations of the processor and RAM sockets.

Notice that there are four RAM sockets. Your motherboard may have a different number of RAM sockets. The RAM sockets are often ordered—you must use each socket in a specific order or the motherboard won't recognize the RAM you add. The manual will tell you which order to use when adding RAM when there is a specific order to follow.

Adding Cooling to the Processor

Before you can put the processor on the motherboard, you need to provide cooling for it. Cooling is added as a fan or a liquid cooling mechanism. Without cooling, the processor will quickly overheat, burn up, and become useless. If you don't have a motherboard that includes heat sensors, the processor can easily burn up before you realize what is happening. Even with heat sensors, you must

FIGURE 5-3 The motherboard has specific locations for processor and RAM.

act quickly to save the processor. Of course, the best course of action is to provide the right sort of cooling so the processor doesn't overheat in the first place. The following sections provide you with an overview and common set of steps to perform the task.

Understanding the Need for Thermal Paste

Your processor is precisely made, as is the cooling mechanism you choose. However, even though they're precisely made, they won't match up exactly. In addition, the various materials used in constructing both processor and cooling mechanism expand by different amounts as they heat and cool. The result is small gaps between the processor and the cooling mechanism that keep the cooling mechanism from doing its job.

These gaps create hot spots on the processor—areas that are hotter than the surrounding areas. Hot spots are bad for the processor because they can cause heating and cooling effects to worsen. In addition, they cause the processor to react erratically and could even result in processor failure.

Thermal paste provides a flexible connection between the processor and the cooling mechanism so that they're in constant contact, no matter what the conditions might be. The thermal paste may look like grease to you, but it's most definitely not grease. The paste is specially designed to transmit heat especially well so the processor can get rid of excess heat efficiently. Adding cooling to a processor without the thermal paste between the two of them almost guarantees processor failure, so make sure you use thermal paste and apply it according to vendor instructions.

> **NOTE**
>
> *Depending on the thermal paste you use, it can be especially hard to get out of fabrics. Make sure you wear older clothing and wear latex (or similar thin gloves) when applying it. If you get the paste on your clothing, use isopropyl or ethyl alcohol to remove it, and then wash as soon as possible afterward. Do not put untreated clothing into the washer—the thermal paste will melt and get all over the rest of your clothes.*

Working with Standard Processor Cooling

Standard processor cooling is for processors that you don't plan to overclock—run at clock speeds higher than those recommended by the vendor. It consists of a really large heat sink with a fan mounted on top as shown in Figure 5-4. The top of the heat sink has the fan; the bottom is a smooth piece of metal. You should connect the heat sink to the processor using clamps (generally) or screws. In some cases, the heat sink doesn't attach directly to the processor, but attaches to the motherboard instead. The heat sink shown in Figure 5-4 uses clamps and attaches directly to the motherboard.

FIGURE 5-4 A heat sink of the example machine's processor.

> **WARNING**
>
> *Never touch the pins on the processor. Even if you have discharged any static in your body, there is always a small chance that you may not have gotten rid of every last bit of static. In addition, the pins are incredibly fragile. Touching them is simply a bad idea. Always touch your processor by the edges. Make sure you read any other vendor-specific warnings before you remove the processor from its packaging.*

Before you clamp or screw the heat sink to the processor, make sure you apply thermal paste to the bottom of the heat sink. Remove the processor from its packaging, if necessary, touching it only by the edges. Position the heat sink carefully on top of the processor (in some cases, the processor isn't square so you need to orient the heat sink to fit properly on the top of the processor). Clamp or screw the heat sink in place. Carefully put the processor back into the bottom part of its packaging until you're ready to put the processor on the motherboard.

Understanding Liquid Processor Cooling

Working with liquid cooling is much harder than working with a heat sink. Liquid cooling uses a liquid to remove heat from the processor. It works much like the radiator does in your car. The advantage of liquid cooling is that it removes a lot more heat, making it possible to overclock your processor for an increase in speed. Because the process varies by cooling kit, there isn't any way to describe the process for installing a liquid cooling kit in this chapter and have the instructions work for the majority of readers. The best option is to view a video of the process online. For example, you can find the procedure for installing an AsusTek liquid cooling system on an AMD FX processor at http://www.youtube.com/watch?v=HC0u2fy0Poo.

> **WARNING**
>
> *Overclocking a processor is always a dangerous process because you attempt to get more speed from the processor than it was designed to provide. The processor could act erratically at some point or even fail because you have overextended it. In most cases, you don't want to overclock your processor unless you have time and resources to fine-tune the overclocking process and to buy a new processor when you fail. If you do decide to overclock your processor, the video at https://www.youtube.com/watch?v=bpwcQH-AATE and the article at http://www.wikihow.com/Overclock-a-PC provide good starting resources.*

Chapter 5: Adding RAM and Processor

Inserting the Processor

After you have the cooling installed for your processor, you can install the processor on the motherboard. Installation is actually a two-step process. First, you must ensure the processor is oriented correctly. Second, you must place the processor in the socket and lock it down. The following sections tell you about this process in detail.

In order to perform this step, you must remove the motherboard from its box. The motherboard will likely reside in an antistatic bag. Open the bag carefully because you need to reuse it later. Always handle the motherboard carefully. Try to handle it by the edges when possible. Never touch any of the exposed pins if at all possible. Always eliminate any static on your body before touching the motherboard to keep from damaging it.

Orienting the Processor

Orienting the processor is essential. Older processors were often square, had the same number of pins on each side, and could be oriented in one of four ways. The only thing that told you which way to orient the processor was a small indicator in one corner of the processor that you had to match with the corresponding indicator on the socket. The results were often catastrophic. Both the processor and the motherboard would literally blow up when you applied power. Fortunately, processors today have safeguards in place to make it less likely that you'll insert the processor incorrectly. Figure 5-5 shows the pin pattern

FIGURE 5-5 Note the keying used on the processor so you can insert it only one way.

FIGURE 5-6 The socket is also keyed to ensure you can't insert the processor incorrectly.

for the example processor. Notice that some pins are missing, so that there aren't an equal number of pins on each side. This technique is called keying the processor.

The socket shown in Figure 5-6 is also keyed. Notice that some holes are missing, so trying to insert a pin in that location will cause the pin to bend. When inserting the socket, it's best to know precisely how the processor should go into the socket so that you don't bend any pins. Look at your processor now and the associated socket. You should see some form of keying that ensures the processor can only go in one way.

The keying is one reason that you must make absolutely certain that you buy a motherboard that provides the socket required by your processor. Otherwise, you can't insert the processor—it simply won't work and you'll only end up damaging the processor pins.

Locking the Processor Down

Make sure the latch for the processor socket is unlatched. You should be able to see through the holes in the socket quite clearly. If the latch is latched, the holes will be partially or fully obscured. The latch must be up before you try to insert the processor or you'll bend the processor pins. Figure 5-7 shows the latch in the up position for the example system.

Chapter 5: Adding RAM and Processor

FIGURE 5-7 Ensure the latch is in the up position before you insert the processor.

Gently insert the processor into the socket. The insertion process should go easily. If the processor doesn't go easily, recheck the orientation and the latch. Under no circumstances should you ever try to force the processor into place. If you do, you'll bend the pins.

Once the processor is in place, you should be able to move the latch into the locked position. Depending on the design of your system, there may be a lock to keep the latch in place. Make sure you lock the latch in place to keep the processor from slipping out of the motherboard. Figure 5-8 shows the example motherboard with the processor locked in place.

If your cooling fan is of the type that attaches to the motherboard, then you need to clamp it in place after you ensure the processor is properly seated. Make sure you apply thermal paste (as needed) to the top of the processor so that you get a good thermal connection between the processor and the fan. The processor kit normally comes with the thermal paste in a tube, already applied to the top of the processor, or already applied to the bottom of the fan. Figure 5-9 shows the motherboard with the cooling fan clamped down (you can't see the thermal paste because the fan is already in place).

Now you need to power the cooling fan. This fan always attaches to the motherboard as shown in Figure 5-9. The plug on the motherboard goes into the fan socket. If your motherboard comes with appropriate software and monitoring

Figure 5-8 The example motherboard now has the processor latched in place.

Figure 5-9 The cooling fan is clamped in place.

Chapter 5: Adding RAM and Processor

features, it can tell you whether the processor fan ever fails, which allows you to replace the fan before the processor overheats in many cases.

Installing the RAM

Your RAM will usually come as a single-inline memory module (SIMM) or a dual-inline memory module (DIMM). Some systems use other types of RAM, but these are the most common types. Figure 5-2 shows a DIMM for the example system. Your DIMM may look a little different, but it works essentially the same. The following sections describe how to install a SIMM or a DIMM into a system (the installation process is the same, just the configuration of the memory module differs).

Looking at the RAM Sockets

The RAM sockets accept the SIMM or DIMM containing memory for your motherboard. The sockets have latches on either end as shown in Figure 5-10. If you look closely, you can see the pins in the socket that match the pins on the SIMM or DIMM. Notice also that there is an insert that matches the cutout in the SIMM or DIMM card. This combination of cutout and insert keys the SIMM or DIMM so that you can't insert it incorrectly. It's essential to orient the SIMM or DIMM correctly before you insert it into the socket.

Inserting and Securing the RAM

Line up the SIMM or DIMM with the socket and push firmly. As you push, the little latches will move up and eventually engage the card, holding it in place. Do not force the SIMM or DIMM in place, but you do need to press firmly to get the

FIGURE 5-10 The socket is designed to accept the SIMM or DIMM in only one direction.

FIGURE 5-11 An example of a DIMM in place on the motherboard.

latches to engage. Figure 5-11 shows one of the DIMMs in place on the motherboard.

WARNING

Make sure you insert the SIMM or DIMM at a 90-degree angle. If the SIMM or DIMM is angled incorrectly, you could bend some of the pins in the socket and the memory module will never click in place. Press straight down, never at an angle. If the SIMM or DIMM doesn't want to go into the socket, you could have it misaligned or oriented in the wrong direction. Always exercise care when inserting the memory.

When you finish this step, you should put your motherboard back into its antistatic bag and then back into its box. Doing so will help keep the motherboard from being damaged while you perform the preliminary steps in Chapter 6. Remember to handle the motherboard by the edges when moving it back to the antistatic bag.

6

Installing the Motherboard

Chapter 5 showed how to install the RAM and processor on your motherboard. Now it's time to get the motherboard into the case. Of course, before you can just stick the motherboard into the case, you need to configure the case, which includes putting a power supply into it. This chapter helps you perform all three steps so that when you're done, you have a setup that you can actually test. The system won't boot yet, but it will make some beeps that will assure you things are working with the motherboard.

> **NOTE**
>
> *Keep the motherboard in its box until you're actually ready to start working with it. Remember to touch the plugged in power supply to remove any static from your body and then unplug the power supply before you attach it to the motherboard. Exercise care in working with your new system. Take your time so that there is less possibility of accidents.*

Configuring the Case

Your case will come in a box with just about everything put together. However, you'll find a box of parts on the inside that contains screws and other pieces of hardware you need to install components into the case. The case itself will likely require some assembly before you do anything else. For example, you may need to install fans in the front, back, or sides of the case as needed to cool the system. The case could also feature parts used to hold storage devices in place that you'll need to secure before you can do anything with the storage devices. The vendor will likely provide a single sheet of instructions telling you what you need to install before you can start to work with other components.

> **WARNING**
>
> *Be especially careful with plastic parts because you may find them a little more fragile than they first appear. For example, your case may require you to install a plastic part containing the front panel lights and power switch. Try to install this piece just one time because the screws for it will hold best if you screw them in just one time.*

It pays to inventory and review the various pieces of the case. For example, the parts bag will likely contain screws of various types. They're all small and look like they could fit anything, but if you look more closely you'll find that the screws differ in various ways. Some screws will have a hex head and rely on coarse threads. These screws are often used to secure add-on boards. Smaller screws with rounded heads and fine threads are often used to secure storage devices to the case. A special set of screws provides the means to secure the motherboard to the case. Figure 6-1 shows an assortment of common screws; Table 6-1 describes each of the screw types and their purposes.

Installing a Power Supply

Most power supplies are about the same size and they all come with the same set of four holes as shown in Figure 6-2. You need to use hex head screws to install the power supply. There is no need to tighten the screws to the point where something breaks, but you do want them snug to ensure the power supply doesn't vibrate loose.

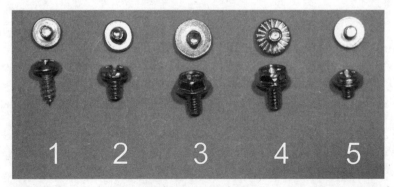

FIGURE 6-1 Each screw type has a particular purpose in putting your machine together.

Chapter 6: Installing the Motherboard

TABLE 6-1 Common Screw Types Used in PCs

Callout	Screw Type	Purpose
1	Flat head, self-tapping	Used to mount fans to the case
2	Round head, 6-32	A Unified Thread Standard (UTS) or United National Coarse (UNC) screw using a number 6 wire size with 32 threads per inch. Used to mount some types of storage devices to the case
3	Hex head, 6-32	Used to mount the power supply to the case and to secure add-on boards
4	Hex head with ridges, 6-32	Used to mount the power supply to the case
5	Round head, M3	A fine-threaded (0.5-mm pitch) 3-mm wire size screw used to mount some types of storage devices to the case

FIGURE 6-2 All power supplies come with the same four hole sets so that they fit any case.

TIP

Your parts bag may come with a special kind of hex head, 6-32 screw that has ridges under the screw head. When you have these kinds of screw available, use them in lieu of a screw that lacks them. The ridges will help the screw stay in place better without overtightening it.

Identifying the Power Plugs and Sockets

The power supply has a number of plug types. One or two of these socket types (depending on the power supply design) go directly to the motherboard and aren't used for any other purpose. A number of sockets are used to power storage devices, peripherals such as fans, and could power a display adapter. They come in two types. Figure 6-3 shows the power supply socket types that you see most often. Table 6-2 tells you the uses for these sockets.

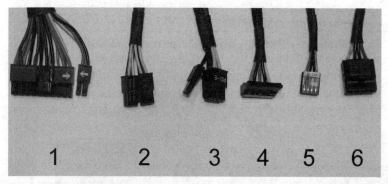

FIGURE 6-3 The power supply provides a number of socket types, each of which has a different purpose.

TABLE 6-2 Common Power Supply Socket Types

Callout	Type	Purpose
1	Motherboard	Connects the motherboard to the power supply. The motherboard, in turn, provides power to each of the add-on boards plugged into it. Most motherboards also supply power to the processor fan.
2	CPU 12-V 4+4 pin	Provides special CPU power as needed. The connector isn't required on most systems.
3	Peripheral Component Interconnect-Extended (PCI-X)	Also known as a Scalable Link Interface (SLI)/Crossfire socket. Connects display adapters that need additional power to provide support for SLI or Crossfire connectivity.
4	Serial Advanced Technology Attachment (SATA)	Connects storage devices that rely on the SATA interface to the power supply.
5	Floppy disk (small device)	Connects storage devices with lighter power requirements to the power supply.
6	Molex (large device)	Connects peripherals such as storage devices and fans to the power supply. You may also need this socket for some types of display adapters.

The motherboard will have a special plug to accommodate the larger socket from the power supply. It pays to know the location of this plug to make it easier to connect the motherboard to the socket after you install the motherboard into the case. Don't plug the motherboard in now—doing so will make it unwieldy to work with the motherboard and make damage more likely.

> **NOTE**
>
> *Depending on your power supply setup and motherboard requirements, you might actually find that you need to attach multiple cables from the power supply to the motherboard. Take time to actually view the motherboard, associated documentation, and cables before you proceed too far in the installation process.*

When working with the various plugs, you should note how the plug is constructed. Each wire has a separate plastic holder as shown in Figure 6-4. Each holder is molded as part of the plug as a whole. Notice that some holders have a squared top and others have a rounded top. The difference in holder shape is meant to help you insert the power connector correctly. When installing the power connector, the plug should insert with a little resistance, but you shouldn't have to push too hard. If you find that you're pushing relatively hard to insert the plug, look carefully at the holder shapes again to ensure you're inserting the plug in the right place and in the right direction.

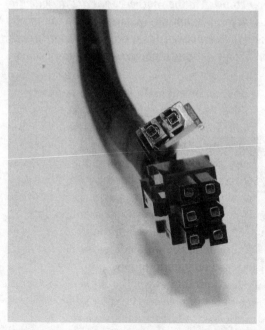

FIGURE 6-4   Look carefully at the wire holder shapes.

Setting Up the Motherboard

Older motherboards had a lot of configuration requirements. Usually you used jumpers to configure the motherboard for various needs. Newer motherboards use few, if any, jumpers. However, it pays to review the documentation to ensure you understand any configuration requirements. For example, when configuring a motherboard for one or more Scalable Link Interface (SLI) display adapters, you may have to change some of the physical configuration.

While you have the motherboard outside the case, it's a good idea to identify other places where you may have to insert plugs or other items. For example, the motherboard accepts all of the front panel connectors and you need to know where to place them. Figure 6-5 shows the position of the front panel connectors for the example system. Your system will be different from the one shown, but all motherboards provide connectors of this sort.

The motherboard will also have special fan connectors. You must locate the connectors because you must connect the processor fan (as a minimum) to the correct connector on the motherboard. It's also possible to connect other fans. These case fans are usually used to cool specific items and the use of the motherboard connector allows you to monitor them. Figure 6-5 also shows the fan connectors for the example system. As you can see, there is more than one of them.

> **NOTE**
>
> *Make sure the fans you connect to the motherboard are designed for that purpose. Don't use a fan without the required connector. In addition, the fan must meet the power requirements for use with motherboard connectors. If your fan draws too much power, it could damage the motherboard.*

FIGURE 6-5 Locate the position of connectors you need to work with.

Chapter 6: Installing the Motherboard

The power supply connectors also appear in Figure 6-5. Depending on your motherboard setup, the power supply may actually rely on more than one connector. It may have a large main connector, with several smaller connectors. Read the vendor documentation to ensure you understand precisely which power connectors to use and when to use them. Some connectors may only be used when you use certain motherboard features.

If your motherboard comes with a special panel for the connectors on the back of the motherboard, you need to install the panel before you install the motherboard. Review the vendor's instructions for placement of this panel—it varies by vendor. The panel provides cutouts for the connectors and provides a nice back-of-case appearance. Most importantly, the panel helps maintain the correct airflow within the case. If you don't have the panel in place, things could fall into the case and some components could overheat. Always install the panel before proceeding if one comes with your system.

Ensuring the Motherboard Is in Place

After you have the power supply installed, the next critical piece of equipment to install is the motherboard. Everything attaches to the motherboard, so ensuring this part is precisely and correctly placed is essential. Missing a standoff (a plastic piece that keeps the motherboard from touching the case) could place undo strain on the motherboard or cause it to short. Placing the motherboard incorrectly could cause problems when you try to install add-on boards. Not screwing the motherboard down correctly could cause the ground plane to function incorrectly, causing intermittent errors. In addition, the motherboard could shift in the case, causing all kinds of transient problems. The following sections discuss all these issues and more.

> ### WARNING
> *Failing to ground yourself before working with the motherboard could cause a discharge of static electricity that will likely cause motherboard damage. It's always better to ground yourself before touching the motherboard. Avoid touching pins or other places where contact could cause a discharge. Obviously, you can't avoid all contact with the motherboard elements, but reducing this contact improves the chances of a good motherboard installation.*

Installing the Standoffs

The standoffs provide support for the motherboard and keep powered sections from contacting the case. A standoff has a definite top and bottom as shown in Figure 6-6. The top has a kind of spring-loaded catch for keeping the standoff

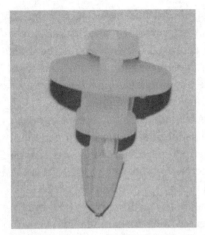

FIGURE 6-6 A standoff is specifically designed to provide good motherboard support.

in place once you insert it into the motherboard. The bottom half is meant to slide into the case slots. The motherboard has holes in it that will match up with slots in the case. Figure 6-7 shows the case slots and Figure 6-8 shows the matching holes in the motherboard.

Place each standoff in its associated hole in the motherboard. If you make a mistake by placing a standoff in the wrong hole, use a needle nose plier to temporarily push the upper portion of the standoff together so that you can push

FIGURE 6-7 The case slots hold the bottom of the standoff.

Chapter 6: Installing the Motherboard

FIGURE 6-8 The motherboard holes hold the top of the standoff.

the standoff back through the hole without damaging the motherboard. The case vendor or motherboard vendor will usually supply more standoffs than you need, so don't worry if you don't use all of the standoffs.

Positioning the Motherboard

After you get all of the standoffs in place and you're sure they're locked in position, you can position the motherboard in the case. The slots and standoffs are designed with some slack so that you can position the motherboard for best reception of add-on boards and other computer features.

Slide the motherboard in place, making absolutely sure that the groove in the bottom of each standoff slides into place in a case slot (not above or below the slot). The motherboard should slide easily—never force the motherboard into place. Moving the motherboard in place will rely heavily on your sense of touch—you won't be able to visually verify that every standoff is precisely in place, but you can feel if something is amiss.

When the motherboard is correctly positioned, there will be a small space between the back of the motherboard and the case as shown in Figure 6-9. You should be able to see the small slots used to hold the add-on boards in

FIGURE 6-9 Ensure your add-on boards will install correctly by positioning the motherboard carefully.

place, but there shouldn't be a large distance between the motherboard and the case. Otherwise, the add-on boards won't fit correctly.

Screwing the Motherboard in Place

The case has special standoff-like protrusions with threaded holes in them as shown in Figure 6-7. The motherboard has associated plated holes in it that match up with these threaded protrusions. When you screw the motherboard in place, you not only physically secure the motherboard, but also provide a connection from the ground plane of the motherboard to the case, so that the motherboard has the ground needed to reduce electromagnetic interference (EMI) and to allow the components to work correctly.

Somewhere between two and four motherboard holes match up with the threaded protrusions in the case. You use hex head screws to make the connection. In some cases, there are special washers or other means to ensure that once you screw the motherboard down, that screw doesn't become loose through vibration. Screw the motherboard down carefully, ensuring that it doesn't shift position as you do so.

Chapter 6: Installing the Motherboard

> **WARNING**
>
> *Don't overtighten the screws. Doing so could cause damage to the motherboard. The motherboard isn't so large or heavy that you need to torque the screws down so they can never be moved again. If nothing else, finger tighten the screw and then use a #2 Phillips to tighten the screw ¼ turn more.*

Connecting the Case Features to the Motherboard

The front of your case will have a number of features associated with it. These features provide you with information and make it possible to connect external devices without having to access the back or inside of the case. Figure 6-10 shows a typical assortment of case features, but your case will likely have a different layout and could have different features associated with it.

The features that your case provides may not match with that of the motherboard. For example, your motherboard may not provide the capability to

FIGURE 6-10 Cases contain a number of useful features that you can use to monitor the system and connect to it.

connect a storage device (SD) card. Even if your case provides the feature, the motherboard won't support it, so the case feature won't be active. Here are the most common case features found on every modern motherboard:

- Power light: Indicates that the system has power applied to it.
- Hard drive light: Indicates that one or more of the hard drives are active. The light doesn't indicate which hard drive and may not light at all when using some host adapter solutions.
- Power switch: Tells the motherboard to apply power to the system or to remove power from the system.
- Reset switch: Supplies the means to return the system to a known state after a system freeze when pressed.
- Universal Serial Bus (USB) connector: Provides access to one or more USB hubs on the motherboard.
- Headphone jack: Allows use of an external headphone in place of the system speakers.
- Mic jack: Allows the user to connect a microphone to the system for recording or speech input purposes.

Depending on your motherboard, you could have other features available, such as an SD card jack. It's important to match the motherboard to the case if at all possible so that the case doesn't end up with a number of nonfunctional features. The following sections describe how to connect the case features to the motherboard.

Identifying the Pins on the Motherboard

The pins to connect the case features to the motherboard are usually in one place. The motherboard vendor wants to make them easy to find. Figure 6-11 shows where the pins are located on the example system. Most motherboards will have the pins in a similar location.

Note that the pins normally come on a header that may contain multiple case features in one group. A close-up of the pins shown in Figure 6-11 appears in Figure 6-12. Notice how the single header contains groups of pins for multiple case features. The writing on the motherboard generally tells you which pins are part of specific features.

> ***WARNING***
>
> *Pay special attention to precisely how the pins are grouped. Connecting a case plug across multiple features can cause motherboard damage. It's important to verify that the plugs are connected correctly before you apply system power.*

Chapter 6: Installing the Motherboard

FIGURE 6-11 Locate the case feature pins on the motherboard before attempting to connect the case connectors.

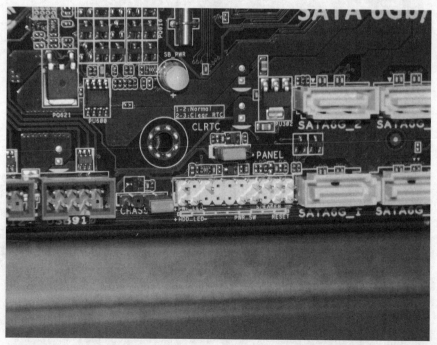

FIGURE 6-12 A close-up shows that the case features are grouped on one or more headers.

Connecting the Case Plugs

The case plugs will be attached to cables (or grouped wires) that come from the front of the case. Each wiring group will be labeled so that you know which feature they access. Figure 6-13 shows a typical grouping of cables.

Carefully plug each cable into its associated set of pins on the motherboard. You may not find that some case plugs have no associated set of pins. Many of the case plugs will plug in immediately next to each other. There won't be any space between the plugs as shown in Figure 6-14.

> **WARNING**
>
> *The motherboard pins bend easily. Don't force the case plug. If it won't fit, verify that the case plug is oriented correctly and that you have placed it directly on top of the pins. The plug should slide onto the pins with little effort.*

FIGURE 6-13   A typical group of case feature cables.

FIGURE 6-14   The case plugs may fit right next to each other on the motherboard.

Testing Your Initial Setup

Your system doesn't have a lot of features as of yet and most certainly won't boot without a display adapter, keyboard, and storage device. However, you can still test your setup to ensure what you have done so far works. The following steps will help you perform the required first-check:

1. Plug the power supply into a surge suppressor.
2. Press the power switch on the front of the case. The power supply fan should start and you may see a light glow on the motherboard. You should also see the power light on the front of the case glow.
3. Listen for the system to beep. The beep is likely telling you there is no keyboard. You could possibly hear a voice telling you that something is missing as well. For now, it doesn't matter what you hear, but you should hear something.
4. Press the power switch on the front of the case. Everything should turn off.
5. Unplug the power supply from the surge suppressor.

7
Providing Video

The display adapter provides video output based on a combination of data sources that include the operating system, applications, and even built-in routines provided with the motherboard. Some motherboards come with a built-in display adapter that you can use for video purpose. In general, the motherboard display adapter will be good enough for business purposes, a security system, or other non-display-intensive uses. However, you do need a high-end display adapter for gaming, drawing, photographic, or other display-intensive purposes. This chapter describes how to install a typical display adapter. Even though the procedure for your display adapter may vary slightly, you'll be able to use the instructions for most purposes.

Understanding the Video Basics

Before you can see anything on your computer monitor or other output device, you need to install a display adapter (assuming your system doesn't come with one installed or you choose not to use the pre-installed display adapter). The following sections describe a basic installation case. Make sure you check your vendor documentation for any differences in your installation.

> **NOTE**
>
> *The photographs in this chapter show the example system. Even though the basic installation procedure is the same for all display adapters, your actual display adapter can vary significantly in appearance from the one shown. Focus on the procedure, rather than on a precise presentation of the display adapter.*

Understanding How Things Work

The display adapter receives video data instructions from the system and any applications designed to work with it. These instructions are interpreted, computations may be performed, objects are created, and then everything is rendered for display on screen. The precise form of rendering depends on the output chosen by the user.

Displays are created using pixels. Looking at the display just from a pixel perspective, a 1280 × 1024 display refreshes 1,310,720 pixels every second. That computing requirement doesn't even consider the work required to create the display in the first place. In order to create the information quickly enough, a display adapter has a graphics processing unit (GPU), special memory, and a high-speed connection to the motherboard. Unlike most add-on boards, a display adapter will generally fit in just one slot. Because display adapters run so hot, they require fans and the use of exotic cooling technology to continue working. In short, the video output represents an incredibly complex process.

At one time, display adapters created analog signals exclusively, but the output of choice today is some form of digital signal because analog data conversion tends to cause data loss (creating a less precise rendering). The formatting of that signal depends on the standard chosen, which is dictated by the device you're using. Most monitors rely on High-Definition Multimedia Interface (HDMI) (see http://hometheater.about.com/od/hometheatervideobasics/qt/hdmifacts.htm) or Digital Video Interface (DVI), which provides a combination of digital and analog signals (see http://homeavcables.com/dvioverview.html).

The capabilities of your display adapter determine the kinds of devices you can attach to it. However, the installation procedure for display adapters is essentially the same, no matter which display adapter you use.

Using a Special Motherboard Socket

The motherboard will contain some number of special Peripheral Component Interconnect Express (PCIe) slots. The lowest version number you want is 2.0. The slots are rated by data rate, with a ×1 slot being the slowest and a ×16 slot being the fastest. The ×16 slot is also the longest of the slots. Figure 7-1 shows an example of two PCIe 2.0 ×16 slots.

In this case (even though you can't see it in the printed version of the book), one slot is colored blue and the other is black. The color difference is important. The blue slot runs at the full data rate. The block slot is sized as a ×16 slot, but it actually runs at the ×4 speed. When selecting a slot for your display adapter, you want to use the full speed slot first. If you later add a second display adapter, you can use the slower slot.

Depending on your motherboard, you could have one, two, or four slots. Each slot will support one display adapter. When configured as a CrossFire or Scalable

Chapter 7: Providing Video   99

Figure 7-1   The PCIe slots look different from other add-on slots on the motherboard.

Link Interface (SLI) setup, the display adapters will work in tandem to produce the images. What this means is that your really fast display adapter will become even faster. When there is more than one slot, you always have to read your documentation to ensure you use the slots in the correct order or the performance will be affected (assuming the motherboard recognizes the display adapter at all).

Finding the Special Power Supply Connection

High-end display adapters consume a lot of power—more than the motherboard can support without help. A CrossFire or SLI compatible power supply will provide a number of special connectors specifically designed for use with the display adapter. Figure 7-2 shows an example of these connectors. In some cases, the connectors will also appear in a special color (with red being the most common). After you install the display adapter, you must power it by connecting one of these direct power connections to it.

Viewing the Back of the Card

The back of the display adapter contains the connectors you use to create a connection to the output device. Most of the time, the output device will be

FIGURE 7-2 A CrossFire or SLI power supply will provide special display adapter connectors.

a monitor or television. However, it could also be a DVD or other recorder, or some other device. Figure 7-3 shows a typical example of a display adapter that supports two monitors as output. The features your display adapter supports depend on its capabilities.

In some cases, the back of the display adapter could include inputs as well. The most common input is the jack used to support broadcast signals of

FIGURE 7-3 A typical display adapter that supports two monitors as output.

Chapter 7: Providing Video

FIGURE 7-4 Some display adapters require input to provide full functionality.

various sorts. If your display adapter has a tuner, it also needs input to support the tuner. This input can take a number of forms, but the most common form is a jack that accepts a coaxial cable as shown in Figure 7-4.

Installing the Video Card Correctly

Most add-on boards you install in the system are simply pushed straight down into the slot and then secured with a screw to the case. Look carefully at the right bottom corner of your display adapter. There is a hook on it, as shown in Figure 7-5. This hook helps secure the front of the card so that the display adapter doesn't have a tendency to unplug itself from the slot when someone adds or removes devices from the back of the card.

> **NOTE**
>
> *In some cases, you must remove the hex head screw in the case that is used to secure the display adapter in place. Remove the screw that is associated with the display adapter slot and set it aside. Don't lose the screw, you need it later to secure the display adapter.*

Before you put the display adapter into the case, you must set the latch at the end of the slot to accept the hook on the display adapter as shown in Figure 7-6. Notice the position of the latch in the figure—it points toward the front of the case.

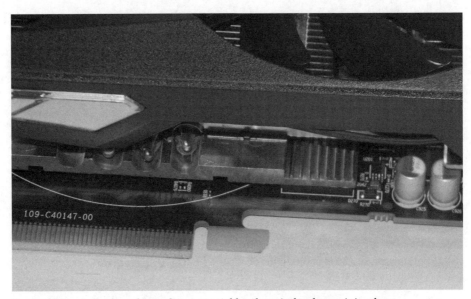

FIGURE 7-5 A display adapter has a special hook on it that keeps it in place.

FIGURE 7-6 The slot possesses a latch that will grab and hold the hook on the display adapter.

Chapter 7: Providing Video

When you push the display adapter down into the slot, the hook will automatically push down at the end of the slot in such a way as to engage the hook. It's very hard to see, but you can feel the latching occur as you push down.

> ### WARNING
> *Pushing down too hard on the display adapter to engage the hook can cause motherboard damage. Flexing the motherboard could cause it to delaminate, creating permanent or intermittent failures that are hard or impossible to diagnose (let alone fix). If necessary, use a nonmetallic screwdriver to help the latch engage the hook on the display adapter.*

After you push the display adapter firmly in place, you will notice the little slot at the top of the back of the card matches up with the screw hole in the case. Figure 7-7 shows a typical example of what you see when this happens. Secure the display adapter using a hex head screw designed for use with the case.

FIGURE 7-7 Secure the display adapter in place using a screw that is designed for use with the case.

Connecting Any Required Cables

Most modern display adapters require that you connect internal cables to them. However, there are still some cases when you don't. Make sure you check the vendor documentation to determine whether you need to connect cables to obtain full functionality or not.

The most common cable you need to connect to the display adapter is the power cable. A high-performance display adapter requires additional power in order to work as anticipated. Most display adapters place the connector for the power cable at the front of the display adapter, near the top, so that the connector is easy to reach. Make sure you connect the power cable before you apply power to the system to prevent potential problems with the motherboard or a loss of display adapter speed. Figure 7-8 shows a typical example of how the cable looks when attached.

Another common cable requirement is to attach the tuner to the sound system in some manner so that you can hear sound from broadcast channels. In some cases, you perform the task directly by attaching the cable from the display adapter directly to the sound card or the sound attachment on the motherboard.

FIGURE 7-8 Connect the power connector to your display adapter.

Chapter 7: Providing Video

> **NOTE**
>
> *The display adapter may not come with the required cable and you can be certain that the motherboard won't supply it either. The vendor documentation will tell you precisely what cable to use in most cases. These cables can take a number of forms. However, the two most common types are shown at http://www.amazon.com/exec/obidos/ASIN/B006JW6IAE/datacservip0f-20/ and http://www.amazon.com/exec/obidos/ASIN/B0028AZ0R8/datacservip0f-20/. The display adapter will still work without the cable, but you won't hear any sound.*

Getting Television Reception Without a Special Display Adapter

Some display adapters do come with a television tuner built-in. These multipurpose display adapters normally give up some speed or other functionality to include the tuner. In many cases, the multipurpose display adapter does everything required of it. For example, in the classroom or as part of a business presentation setting, you really don't need anything else. This setup will even work for home entertainment center purposes as long as the games you play aren't too challenging.

Unfortunately, the loss of speed becomes noticeable when the user wants to play high-end games or perform tasks such as use a computer-aided design (CAD) application. In this case, you can add a second adapter to your system that contains just the tuner. For example, the Hauppauge 1196 WinTV-HVR-1250 PCI-X ×1 TV Tuner 1196 (http://www.amazon.com/exec/obidos/ASIN/B0014YFC18/datacservip0f-20/) requires a display adapter slot, but you could use the PCIe slot in a system to support it (relying on the second or slower slot to provide the display adapter with a higher-speed data connection).

The tradeoff in this setup is that if you have an SLI system and only two slots, you give up the ability to add a second display adapter to speed the rendering of graphics. In many cases, this won't be a problem. However, you could notice the performance drop when playing certain games that require extensive processing power.

Considering CrossFire and SLI Configuration Needs

Systems that have CrossFire and SLI support let you combine the power of multiple display adapters to create an extremely fast setup. However, these systems are also designed to use a single display adapter, which is how they're usually configured at the factory. It's far more likely that a system will have a single display adapter, than two or four display adapters. In order to use the CrossFire or SLI functionality to gang multiple display adapters together, you must perform some configuration tasks. The following sections describe the process for creating a CrossFire or SLI configuration with your system.

> **NOTE**
>
> *This book assumes that most people will use two display adapters. Four-display-adapter systems are relatively rare and you need a special motherboard to create one. However, the process is the same whether you use two or four display adapters—the major consideration is that when working with four display adapters you must perform the configuration process for each additional display adapter you want to use.*

Configuring the Motherboard

Depending on how your motherboard is put together, you could see a variety of methods for connecting the two display adapters together at the motherboard level. The example system used a set of headers on the display adapter as shown in Figure 7-9. The figure shows the cable required to create an SLI configuration in place. This setup has the advantage of not getting in the way of the display adapters unless you have particularly fat display adapters (usually of the double-slot variety).

FIGURE 7-9 The two headers provide a connection for the display adapter.

> **NOTE**
>
> The SLI connector cable will likely come with your display adapter, and not with the motherboard. If you don't receive a cable with either the display adapter or the motherboard, you need to contact the vendors to obtain the required connector. It's also possible to buy a cable separately, such as the one shown at http://www.amazon.com/exec/obidos/ASIN/B0025VT6LQ/datacservip0f-20/, but you must ensure the cable will actually fit your system and provide the required connectivity.

In order to use this setup, you place a jumper between the two headers on the display adapters. The vendor supplies the jumper for you. Exercise extreme care in putting the jumper in because it tends to deform easily or you could damage the header in some way. The jumper should insert relatively easily.

> **WARNING**
>
> Make sure you take precautions for static electricity before you put the jumper cable in place. Touching the display adapter pins could cause damage to the display adapter electronics if you release static electricity.

Using the Correct Slot

Many motherboards that have multiple PCIe slots have preferential slots that run at full speed. In order to make your display adapter work correctly, you must use this slot first. It's not just a matter of speed (although getting the best possible speed is always a good reason to choose the correct slot), it's also a matter of usage. The system will only enable one monitor during the boot process. This monitor is the first output of the first slot—the preferential slot. All the other monitors will remain darkened until after the system boots and the operating system activates them.

The slots are generally color coded. Of course, if you're color blind, the color coding won't help much. The trick here is to put in just one display adapter, run the procedure found in the "Performing a Quick Video Test" section of this chapter, and if you don't see output on your monitor, you need to try the other slot. The preferential slot is normally the one closest to the power supply. This rule of thumb works for the vast majority of motherboards. Figure 7-1 shows an example of a system with two PCIe slots.

In general, you use two of the same display adapter when creating a CrossFire or SLI setup. However, there are cases where you can mix different, but compatible, display adapters and get a working solution. This second case requires careful selection of display adapters and using the correct display adapter in each slot. The fastest of the two display adapters should always reside in the slot with the fastest data bus to improve overall system performance.

Making the Required Power Supply Connections

The power supply provides more than one circuit to support the various power needs of the system. Look carefully at the bundles coming out of the power supply. Each bundle represents a different circuit in most cases. The display adapters you connect to the power supply create a major drain on the circuit. Whenever possible, use connectors from different bundles for each of your display adapters to distribute the load better and reduce the probability of power supply failure.

Of course, each display adapter will require its own power supply connection. As a result, you need a power supply that can provide enough of the right connections to support each display adapter. Most CrossFire and SLI compatible power supplies provide two such connections. Make sure you determine the number of available connections before you buy the power supply.

Making Connections Between Cards

The motherboard connection you created earlier connects the data bus between the two cards. However, the cards may also require synchronization of other signals. If this is the case, the card will come with a special set of connectors at the top of the board. You connect the two cards using a special jumper.

> **NOTE**
>
> *The jumper cable may actually be an extra item. Make sure you check for the need of a jumper cable as part of your display adapter purchase. Ensure the cable is provided with the display adapter or buy the required extra cable from the vendor.*

Considering TV Tuner Configuration Needs

A display adapter with a television tuner (or a separate card used as a television cable) requires an input connector. In addition, there is an audio connection to the sound card (or built-in sound system). These two connection requirements are normally all you need to worry about.

In addition to the connections, television tuners may require that you perform special configurations to ensure that there are no electromagnetic interference (EMI) issues. For example, you may have to attach a block to the input cable to reduce the chance of EMI causing problems for other connections in your area. Make sure you read the documentation carefully to ensure you address all of the EMI requirements for the television tuner.

Although it's rare, a television tuner may require a special power supply connection, much as high-end display adapters do. If this is the case, make sure

Chapter 7: Providing Video

you have a power supply connection of the right type for the television tuner. Because a television tuner does use less power in most cases than a high-end display adapter, you may not need an actual CrossFire or SLI power supply connection to support it.

The television functionality of your system won't be usable until you install the special software required to support it. In fact, you shouldn't be surprised if the system refuses to recognize the television tuner until such time as you install the required operating system drivers. A set of drivers will come with the television tuner card, but the best idea is to download the latest drivers directly from the vendor site. Doing so ensures that the television tuner will work as expected.

In addition to drivers, you also need special applications to work with a television tuner. The vendor will likely supply an application to use. However, you may find better third-party products for your television tuner by looking online. The application you use depends in large part on your personal requirements for using the television tuner.

Connecting Devices

The devices you connect to your display adapter will vary depending on your output needs. However, it's always a bad idea to connect or disconnect a device while the system is running because the electricity could arc from the connector on the display adapter to the plug on the device—damaging the display adapter, the device, or potentially both. Here are some of the devices you might plug into your display adapter:

- Monitor: The most common connection for your display adapter is going to be a monitor. A monitor differs from other sorts of devices in that it doesn't usually provide a tuner or other internal data source. In addition, it can't manipulate the information it receives in any substantial manner. Better quality monitors can accept both analog and digital input. If this is the case with your monitor, the monitor will normally display messages that show it's switching between analog and digital mode until it finds a signal to use.
- Television: A television can act as a monitor. However, a television uses a different aspect ratio than a monitor does in most cases, so the television output will appear differently than the monitor output. In addition, televisions can accept myriad other data sources and often includes an internal tuner or other data source.
- Recording device: It's possible to attach a recording device to your display adapter, just as you would when using other data sources. If your recording device includes a passthrough or output connector, you can still attach a monitor or television to the recording device output.

- Projection system: A projection system makes it possible to display a presentation to a group. On display adapters that have multiple outputs, it's often possible to have the same presentation on a monitor for the presenter and on the projection system for the group.
- Other: A display adapter could be used to power other sorts of devices. The essential consideration is whether the display adapter provides the correct output connector for the device you want to connect.

> **NOTE**
>
> There are adapters to allow use of devices that your display adapter might not normally support. For example, if you have a device that has a Video Graphics Array (VGA), analog, and plug, but your display adapter only has a Digital Visual Interface (DVI) connector, you can use a DVI-to-VGA adapter to make the connection (such as the one shown at http://www.amazon.com/exec/obidos/ASIN/B004G3WCLM/datacservip0f-20/). However, you must have a DVI-A (analog only) or DVI-I (integrated, both analog and digital) connector to make this solution work. If you have a DVI-D (digital only) connector on your display adapter, then you can't connect it to a VGA device, even if you have the DVI-to-VGA adapter.

Display adapters can be configured in several different ways. The most common method is to provide two DVI-I connectors on the back. However, you might find combinations of DVI-I and HDMI connectors. The example display adapter actually contains four ports on the back: DVI-I, DVI-D, HDMI, and DisplayPort as shown in Figure 7-10.

> **NOTE**
>
> DisplayPort display adapters can transmit both video and audio signals over a single cable using packetized transmission (akin to Ethernet). You can use special adapters to convert DisplayPort signals to DVI-I, HDMI, or VGA signals. There are also three-in-one converters like the one shown at http://www.amazon.com/exec/obidos/ASIN/B005H3Q56I/datacservip0f-20/. Many display-adapter vendors use the DisplayPort output to transmit 3D stereoscopic signals to an output device such as a television or specially equipped monitor.

It's important to know how to differentiate between various kinds of DVI connectors. Figure 7-11 shows the various DVI connector types as line drawings. Notice how the various connector variations make it easier to determine precisely what sort of connector (or plug) you have at your disposal.

Chapter 7: Providing Video

FIGURE 7-10 Display adapters can include a number of connector types, such as the configuration shown here.

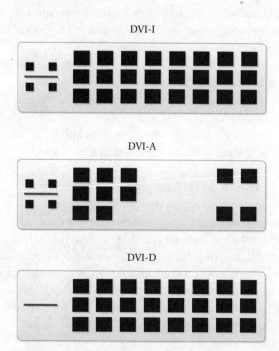

FIGURE 7-11 Differentiating between the various DVI connector types is important for determining how to configure your system.

Performing a Quick Video Test

After you have installed the display adapter it's a good idea to test your setup. Your system still won't boot properly, but you should be able to see enough information on screen to ensure your display adapter, the cable connecting the monitor to the system, and the monitor itself are all working as expected. The following procedure will help you test your setup:

1. Check that the power connector, if one is needed, is properly connected to the display adapter.
2. Verify that the display adapter is properly seated in the motherboard slot and screwed into place.
3. Ensure that any ancillary internal cables (such as an audio cable) are connected properly.
4. Verify that you have at least one output device, such as a monitor, attached to the primary (normally the top or left) connector of the display adapter. (Remember that secondary outputs won't be active during this test because they normally require operating system support.) Try not to use a device that requires an adapter for this test because doing so will add another potential failure point, making troubleshooting far more difficult.
5. Check any input cables, such as a coaxial cable used for television signals, are connected. The inputs won't work at this stage, but it's a good idea to verify that connecting them doesn't cause problems.
6. Plug in the computer and apply power to the system. Verify that the various computer fans have started (there should be fans for both the power supply and the processor as a minimum, but your system could contain additional fans).
7. Verify that the display adapter fans start when power is applied. If the fans don't start, shut the system back down and check for a special power connector for the fans.
8. Perform steps 6 and 7 as needed to ensure proper fan operation.
9. Allow the power to cycle long enough so the display adapter becomes active and shows information on the display. The type of information may not be usable or significant, but you should see something. If you don't see any information, you may have used the wrong connector on the back of the display adapter. As an alternative, the monitor may be set to the wrong input type. Automatic cycling between digital and analog output may not work as expected. Try setting the monitor for a specific kind of input (as supplied by your display adapter output).

Chapter 7: Providing Video 113

10. Perform step 9 until you see output from the display adapter or determine that the display adapter, cable, or monitor is faulty. Any output you see will likely be white text on a black background. If you see odd colors, striations, or colored dots on screen, it could indicate that the monitor is failing or that the cable isn't connected properly.

> **NOTE**
>
> *The system won't boot, so you'll likely see the partial results of a power on startup test (POST) and then an error message. The error message will likely tell you that you're missing a boot device, keyboard, mouse, or some other component that you haven't installed yet. The only purpose of the test is to ensure that you can obtain some sort of output from your system so that you can more easily troubleshoot your system in future chapters.*

8
Mounting Permanent Storage

The need to store data permanently has been evident since the beginning of modern computers. At one time, computer users relied on paper tape, punch cards, and magnetic tape. Of course, newer computers began using ever larger random access hard drives because they provide such a speed boost and the form factor is incredibly small compared to some other storage techniques. Modern computers rely on solid-state drives (SSDs) that are basically a special kind of RAM that emulates a hard drive and there is also optical storage. Future computers may rely on exotic techniques such as holographic data storage (see the article at http://www.techradar.com/us/news/computing-components/storage/whatever-happened-to-holographic-storage-1099304 for details).

> **NOTE**
>
> *For the purpose of this book, hard drives and SSDs are treated equally because the techniques for working with them are the same. Whenever you see a hard drive technique, you can also use it with an SSD.*

From a computer building perspective, it doesn't seem to matter which kind of permanent storage you use. There are basic techniques for installing and accessing data storage that don't really rely too heavily on the kind of storage. Most permanent storage requires that you provide power and a data cable so that the system can access and interact with the data. Yes, there are exceptions. For example, a thumb drive generally attaches to your Universal Serial Bus (USB) port and you remove it when you're done working with the data it contains. There isn't a power connector because the device power comes directly through the USB port.

Some storage solutions can require a third cable. For example, when working with optical drives, you might need to connect an audio cable to provide a path for sound data to the audio card (allowing playback of CDs and DVDs).

This chapter explores all sorts of permanent storage options. However, it pays to remember that the techniques shown here haven't changed all that much, even though devices today can store vast amounts of information compared to devices of old. The techniques you find here will likely work well with newer devices as they come along.

Understanding Hot Swap Functionality

Early computers always required that you shut them down, perform any permanent storage changes, and then reapply power before the new device could be recognized. In addition, you had to reconfigure the system every time you made a change. However, businesses can't run well with systems that require regular reboots. In fact, most enterprise systems run constantly today because people are always using them (even at night, enterprise systems stay on in order to receive updates). This old technology is called cold-pluggable and is still used when working with most hard drives.

When a system is configured properly, it becomes possible to use devices that are hot swappable, which means you can change them out and the system automatically recognizes the change. The system is configured such that the act of removing an existing device and plugging in a new one doesn't cause any permanent damage. The device and the system are protected from any transient spikes in power. A USB thumb drive is an example of a hot swappable device.

Some hard drives can be made hot swappable. However, you need a special case, special drives, and special software to make this all work. The case usually provides front panel access to the drives so that you can remove the drive without opening the case. The power and data cables are made longer so that you can remove the drive from the front and reattach a new drive. The drives are designed so that they can slide in and out of the case without problem. The most important consideration is that the drive has electronics that are capable of surviving a hot swap. Software, such as Windows, will recognize a hot swap and reconfigure the system appropriately. Even Windows requires special drivers to make this happen.

Understanding Permanent Storage Basics

It's helpful to understand a few permanent storage basics before you proceed to install a device in your system. For example, not all hard drives come in the same sized case. Some permanent storage solutions don't even reside in the case. Understanding the various options makes it possible for you to understand the

Chapter 8: Mounting Permanent Storage

configuration requirements of any drive that you obtain with greater ease. It also makes it possible for you to make better choices when locating upgrade options. With this in mind, the following sections provide you with details on permanent storage features.

Defining the Form Factors

The form factor determines what you must do in order to use the device. Some form factors are nearly automatic to use, while others require that you open the case, install, and configure them before you can use them. Most permanent storage options come in one of the three form factors:

- Internal drive: An internal drive resides inside the computer case. You attach the drive by putting it into a drive bay and using screws to keep it in place. When a drive is too small to fit in the drive bay, you use an adapter to make the drive take more space so that it fits within the drive bay properly. As part of attaching the drive to the case, you attach a power cable and a data cable so that the device can communicate with the computer. In some cases, you attach the power cable and data cable before you insert the drive into the case—in other cases you perform the task in the reverse order.
- External drive: An external drive is simply an internal drive that is housed in its own case. The device is attached to the system using a connector on the back of the system. In some cases, the single cable provides both data and power functionality, but it's far more likely that you'll need a separate power cord for the external drive. As with an internal drive, you normally need to configure an external drive before you can use it.
- Self-contained: A self-contained device simply plugs into the system. In most cases, the device is ready to use the moment the system sees it. The device also obtains its power through the connector you use to attach it. The USB thumb drive falls into this category.

The technique used to work with storage media also differs between devices. The following list describes some of the storage media options:

- Permanent: A hard drive is an example of a device with permanent media. The media is always the same in a hard drive. The advantage of permanent media is that it can be faster, hold more data, and is often more reliable than other forms of media.
- Non-writeable replaceable media: A drive can accept media that is pre-recorded with data. You can't change the data in any way. Some DVDs and CDs are this way. When the media is pressed at a professional facility, it provides exceptionally long storage times and is relatively reliable.
- Write-once replaceable media: In some cases, a device will allow you to write data to the media one time. At that point, the data becomes

permanent. For example, a DVD-ROM disk provides this type of storage. Even though the data is permanent, it can degrade quickly, becoming unreadable in as little as 5 years. You use this type of media for short-term storage of data you don't want to be changed.

- Read-write replaceable media: Some replaceable media lets you perform both read and write tasks. You can read and write the media just as you do a hard drive. Depending on the media type, you can get a range of results. For example, magnetic tape provides extremely long data storage times, but magnetic tape is notoriously slow. On the other hand, a DVD-RAM disk provides relatively fast read-write operations, but it suffers from a relatively short storage time and you can only read and write the disk a certain number of times before the disk fails to accept any new data.

The kind of drive you have determines precisely how you have to install, configure, and use it. For example, when you install an optical drive in your system, you don't have to provide any media with it. However, you also can't boot off an optical drive unless it does contain media and the system is configured to allow an optical drive boot. Most systems look for some type of hard drive for boot purposes, but allow other storage options. One of the more common alternatives is the use of a thumb drive (USB flash drive) for boot purposes.

Considering the Power Cable

As with most peripheral devices, storage devices require power. However, not all storage devices use the same kind of power connector. There are three basic power connector types:

5.25-inch 4-pin connector: The most commonly used connector type. It's also known as a Molex connector, after the vendor who originally created it. The connector provides a +5-V wire, +12-V wire, and two ground wires. Every power supply made provides Molex connectors, so you never need to worry about getting an adapter for one. However, you may occasionally need more Molex connectors—a need you can handle using a Y-splitter such as the one shown at http://www.amazon.com/exec/obidos/ASIN/B001PI9AAC/datacservip0f-20/.

Chapter 8: Mounting Permanent Storage

> **WARNING**
>
> *When using a splitter, you need to determine whether the power supply can actually handle the additional load on that particular line. Adding a number of splitters of any sort to a single line could overload the power supply and cause it to fail. Limiting additions to a single splitter per line is usually the best idea.*

3.5-inch 4-pin connector: This is a smaller version of the 5.25-inch 4-pin connector that is commonly used for floppy disk and zip drives. Because storage devices that depend on this type of connector are becoming rare, your power supply may not even provide one. Fortunately, you can get an adapter such as the one found at http://www.amazon.com/exec/obidos/ASIN/B0002J1KW6/datacservip0f-20/ to provide the required power connector should you need one.

Serial Advanced Technology Attachment (SATA): Newer storage devices that rely on the SATA standard use this type of power plug. Any new power supply will provide at least a few of these plugs. However, if you run out of SATA plugs, you can always convert a Molex plug to SATA form using an adapter like the one shown at http://www.amazon.com/exec/obidos/ASIN/B00009YFTI/datacservip0f-20/. It's important to note that using an adapter means that your SATA plug will lack a 3.3-V source, which may cause problems with some storage devices.

There are a number of variations of some of these cables and you need to ensure you have the correct power cable on hand. For example, the illustration shows a standard SATA power cable. However, there are actually several SATA power cable configurations:

- Standard: Includes 3-V, 5-V, and 12-V connectors, ground, and a staggered spin-up selector. This is the largest of the three connectors and the one that most drives (including the drives for the example machine) use.

- Slimline: Provides a 5-V power connector, ground, diagnostic signal, and device present signal. This is the kind of power connector generally used for notebook computers and the storage devices supported for them. Because a desktop-PC host adapter lacks the appropriate signals for a notebook storage device, you should avoid getting notebook compatible storage devices for your system (which also means you don't need to worry about this power connector type).
- Micro: Provides 3.3-V, 5-V, and ground connectors. This particular connector is used with SATA 2.6 compatible storage devices that are in the 1.8-inch form factor. You could possibly encounter this kind of drive when working on a desktop system, but you'll have to have an adapter to use it because most power supplies won't supply one for you. You can find such an adapter at http://www.amazon.com/exec/obidos/ASIN/B002P6PBAQ/datacservip0f-20/.

Considering the Data Cable

Every storage device has at least two cables attached to it. The first is the power cable discussed in the previous section. The second is the data cable. There are so many variants of the data cable that it would be impossible to cover them all here. The best solution is to specifically look at the interface supported by your motherboard or add-on host adapter, get a drive that matches it, and use a cable that's appropriate for making a connection between the two. Here are the most common data cable types that you find in PCs.

SATA: The SATA data cable is the most common type in use today. In fact, many systems out there only use SATA cables. There are actually three kinds of SATA cable and you need to ensure you obtain the right type for maximum throughput (using the wrong cable will decrease your data throughput in some cases):

- SATA I: Has a 1.5-Mbps interface and a potential throughput of 1.5 Gbps. The interface bandwidth is 150 MBps.
- SATA II: Has a 3-Mbps interface and a potential throughput of 3 Gbps. The interface bandwidth is 300 MBps.
- SATA III: Has a 6-Mbps interface and a potential throughput of 6 Gbps. The interface bandwidth is 600 MBps.

Chapter 8: Mounting Permanent Storage   121

USB external: You often find USB used in situations where a new system has to support an older drive system. However, even newer drives can rely on USB. Most USB setups are external drives, so you use a standard USB connector. All you need to do to use the drive is plug it into any USB port. The system automatically recognizes the drive and you can start using it as long as your system has an appropriate driver (which it will have in most cases).

USB internal: Your system likely has posts on the motherboard dedicated to internal USB use. The posts reside on a header that acts as a plug. If you have an internal USB drive, you simply connect the drive to the internal plug using the supplied socket. Of course, this action uses up one of your USB hubs and could make it harder to provide external connections, so you need to choose between using the device internally or externally. If you want to convert your internal drive to an external drive, all you need is an enclosure like the one shown at http://www.amazon.com/exec/obidos/ASIN/B00E362W9O/datacservip0f-20/.

Integrated Drive Electronics (IDE): This is an older cable type that transfers data in parallel, rather than serial format, as SATA does. You may find some systems that still rely on the older IDE specification. It's also possible to buy IDE drives from places such as Amazon.com online. So, this drive specification isn't completely finished yet, but it's on the way out. If you do use IDE, make sure you rely on the IDE66 standard because it has become difficult to obtain older cable types. The IDE66 specification provides a bandwidth of 66 MBps, which makes it slower than SATA drives.

Floppy drive: It's impossible to find an off-the-shelf system with a floppy drive any longer. Even ZIP drives are extremely hard to find. Both floppy and ZIP drives used to use some form of the floppy data cable at one time. (Notice that a floppy cable has a twist in the ribbon cable itself before it gets to the plug.) However, newer drives generally rely on USB connectors and you connect them as external devices, rather than internal devices. It's possible to get internal devices as

well. Many of these drives come with an IDE connection, rather than the original floppy connection. You can also find these devices with parallel port connectors and there are even a few that come with Small Computer System Interface (SCSI) connectors. Even with all these options, you may encounter a situation where your new system simply won't work with that old drive that you really do have to support. Fortunately, you also have the option of using converter kits to turn your drive into a USB setup, such as the one shown at http://www.amazon.com/exec/obidos/ASIN/B001OORMVQ/datacservip0f-20/.

When working with SATA, you can use a SATA III cable for a SATA I or SATA II interface and not lose any throughput. Likewise, using a SATA II cable for a SATA I interface won't reduce the throughput. However, if you were to use a SATA I cable for a STAT II or SATA III interface, the throughput would be limited to 1.5 Gbps. Unfortunately, any of the cables will fit with any of the plugs. You must read the text on the cable to determine the cable type (or ensure you order the correct type during a purchase).

You might encounter other cable types. It all depends on your system and the kind of devices you install. In most cases, these other cable types support older storage devices that are outdated or hard to obtain. Sticking with SATA or USB is the best way to go when you can do so because it has the longest life expectancy.

Working with Drive Size Adapters

Drives come in a number of physical sizes. Older drives always consumed a 5.25-inch drive bay because old technology was bulky. As technology improved, the vendors concentrated first on making 5.25-inch drives hold more data, but quickly found they could also shrink the size of the drive. At first, that meant changing the drive height from a full-sized drive to a half-height drive, and still making the drive hold more data. Then the drive size started shrinking. Consequently, 5.25-inch drive bays eventually saw use in holding 3.5-inch drives. Notebooks now support 2.5-inch drives and there are even 1.8-inch drives.

These various form factors all have to fit within a case designed to hold 5.25-inch drives. The reason that cases use the 5.25-inch form factor is that larger drives still come in 5.25-inch sizes and the case must accommodate the larger drives. The way to make smaller drives fit into a larger drive bay is to use a drive adapter. The drive adapter fits within the 5.25-inch bay, but provides support for the smaller drive. Figure 8-1 shows a typical example of a drive adapter for a 3.5-inch drive.

Chapter 8: Mounting Permanent Storage

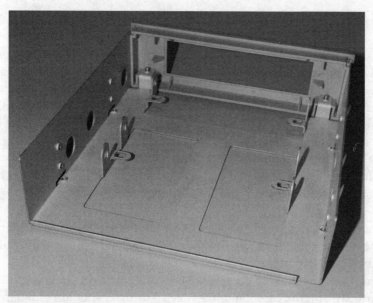

FIGURE 8-1 A 3.5-inch to 5.25-inch drive adapter lets you put a 3.5-inch drive in a 5.25-inch drive bay.

> **NOTE**
>
> *It's currently not possible to adapt a 1.8-inch drive directly to a 5.25-inch bay. You must first convert the 1.8-inch drive to a 2.5-inch drive and then use a second converter for the 5.25-inch bay. However, it's important to verify that the 1.8-inch drive will actually work in your system before you go to the extra effort of adapting it. In most cases, you're better off getting a 2.5-inch or 3.5-inch counterpart for the storage device you want to use.*

Before proceeding to the installation sections of the chapter, make sure each of your drives has the proper drive adapter attached (when the drive needs one). An optical drive will require case front access. A few cases do come with 3.5-inch access plates and associated bays, but most rely on 5.25-inch bays and access plates, which means that you must configure your optical drive for use in a 5.25-inch bay. Most cases do provide 3.5-inch drive bays for hard drives and a few even come with 2.5-inch drive bays. Check your case before you buy and install an adapter to determine whether the drive will fit without one.

Installing a SATA Drive

Your hard drive will likely use a SATA interface, even when you use an SSD drive. Using a bay specifically designed for hard drive use will make installing the hard drive easier and provide better air flow to the hard drive during usage. Because hard drives are in constant use (as contrasted to optical drives, which typically receive intermittent usage), air flow is especially important to keep the drive from overheating and failing. In fact, some builders will actually mount a fan directly next to the hard drive to keep it cool.

The following steps assume that you plan to use a hard drive bay. If you plan to use a front of case bay instead, you switch steps 1 and 2.

1. Attach the power and data cables to the hard drive. You need to attach one of each cable as shown in Figure 8-2. Make sure you position the cables correctly and seat the connectors completely.
2. Slide the drive into the drive bay. Make sure the drive screw aligns correctly with the case mounting holes. The drive shouldn't hinder closing the case front or require the removal of an access panel (which means using the rear screw hold set).
3. Screw the hard drive in place using the round head, 6-32, screws described in Chapter 6 as shown in Figure 8-3. Don't over-tighten the screws. Make sure the screws are snug, but don't place a strain on either the case or the drive assembly.

Figure 8-2 The power and data cables provide a connection between the hard drive and the system.

Chapter 8: Mounting Permanent Storage

FIGURE 8-3 Use the rear set of screws so the hard drive doesn't stick out from the front of the case.

Accessing the Drive Configuration

Whenever you install a new permanent drive using an interface such as SATA, the system recognizes the drive and configures the system setup as appropriate. In days gone by, you had to perform this process manually, but today everything occurs automatically. Of course, automatic configuration only works when you have a default setup, which is the case most of the time. So, you may never need to access the drive configuration, but it's important to know how to do it just in case.

The drive configuration is stored on the motherboard in special permanent memory. In order to access this information, you press a special key during the boot process. Many systems will display the configuration information when you press Delete, but some systems use other keys or key combinations. Check your motherboard documentation to determine precisely which key to press.

The main hard drive information, the drives that the system recognizes, normally appears on the main page or on a special hard drive page of the configuration application. You need to consult your vendor manual to see precisely where this information appears. The hard drive information should match the vendor information for the drive.

The configuration screen is important for another reason. One of the pages, likely a boot configuration page, will contain settings that let you control how the system boots. For example, you could set the system to boot from an optical drive first, a USB drive second, and a hard drive third. Setting the boot cycle in this manner would allow you to use alternative boot media when required. The system searches each location for boot information when the system first starts and uses the boot information in the first available location.

Installing an Optical Drive

Most modern optical drives use a SATA interface. However, you can still find optical drives that use other interface types. Make sure your optical drive uses an interface that the motherboard supports. For the example system, this means using a SATA interface.

Unlike a hard drive, an optical drive requires front of case access so that you can insert and remove media. The following steps help you install an optical drive.

1. Remove a front panel access cover to allow optical drive access. Retain the access panel cover for future use. The access panel cover will simply snap into place in most cases, but a few cases may require you to unscrew it before removing it.
2. Slide the optical drive into place. It helps to slide the optical drive a little further back than necessary to make accessing the plugs easier. If your cables are long enough, you can theoretically attach them before you slide the optical drive in place, but this is seldom the case.
3. Attach the power and data cables to the optical drive. You need to attach one of each cable as shown in Figure 8-2.

FIGURE 8-4 An optical drive will be flush with the front of the case so that you can insert and remove storage.

Chapter 8: Mounting Permanent Storage

4. Align the optical drive with the front set of case screw holes so that the front of the optical drive is flush with the front of the case.
5. Screw the optical drive in place using the round head, 6-32, screws described in Chapter 6 as shown in Figure 8-4. Don't over-tighten the screws. Make sure the screws are snug, but don't place a strain on either the case or the drive assembly.

Working with Solid-State Drives

SSDs and hybrid drives that include both an SSD and a standard hard drive are quite popular now for a variety of reasons. An SSD will install in your system much like any other standard hard drive. However, most SSDs come in a 2.5-inch form factor, rather than a 3.5-inch or 5.25-inch form factor. This means you usually require an adapter when working with an SSD unless your case supports 2.5-inch drive bays.

In general, SSD drives have some significant benefits over using a hard drive. The following list tells you about the pros of using SSD drives:

- Speed: An SSD has no moving parts and can read data at the speed of RAM, making it considerably faster than a hard drive.
- Fragmentation: SSDs don't care where the data appears on the drive because there is no read head to reposition, so even if the data becomes fragmented it's a nonissue.
- Durability: There are no moving parts within an SSD, so it isn't as easy to damage one through vibration or other environmental factors.
- Noise: The lack of moving parts means that an SSD doesn't make any noise.

In order to obtain the benefits of an SSD drive, you must be willing to put up with a few deficiencies as well. The following list tells you about the limiting factors for SSD drives:

- Price: Standard hard drives cost considerably less than SSDs.
- Size: The largest SSDs come in at about 1 TB right now. Hard drives will likely maintain a size advantage over SSDs simply because hard drive technology allows easier size expansions.
- Availability: It's easier to find a standard hard drive in the size and form factor that you want.
- Read and write cycles: An SSD has limited read and write cycles that could lead to early read and write failures. Yes, hard drives wear out too, but hard-drive read and write reliability has improved so much that it's likely the system will be obsolete long before the drive wears out.

Precisely who benefits from an SSD drive depends on how the user employs the system and the storage requirements of any applications. However, these groups are most likely to benefit the most from using SSDs:

- Gamers: This group will likely want a hybrid setup to get the speed advantage of SSD coupled with the storage capacity of a hard drive.
- Scientists and researchers: Performing calculations quickly is an essential part of both science and research. If the data sets are small enough, this group can easily get by with an SSD.
- Entertainment: If you spend a lot of time recording music or performing other tasks that require maximum quietness, an SSD solution is the best option. However, if you want to store vast amounts of data for movies and other home needs, a hybrid setup may work better.

Using USB Storage

Thumb drives (also known as flash drives or USB storage) have an advantage over other solutions mentioned in this chapter because you can simply plug them in with the system running and gain access to portable memory. Figure 8-5 shows

FIGURE 8-5 Plug the thumb drive into the back of the system for semi-permanent storage.

Chapter 8: Mounting Permanent Storage

FIGURE 8-6 Plug the thumb drive into the front of the system for moveable storage.

a thumb drive plugged into the back of a system because it's used to provide semi-permanent storage. Figure 8-6 shows a thumb drive plugged into the front of the system. In this case, the thumb drive provides storage to move data between systems. Where you plug the thumb drive in depends on how you plan to use the storage.

Most operating systems automatically recognize that you have either inserted or removed a USB drive. The small storage capacity offered by USB storage makes it unsuitable as a boot solution (or a main drive) for most systems, but the drives make it possible to move your pictures, videos, and music around with ease.

There are times when it would be nice to create the ultimate in secure systems by using a USB drive as a main drive. That way, at the end of the day you could simply remove the USB drive from the system and store it in a vault somewhere. For example, you might create a system to store confidential data that you can't simply leave in the system overnight due to security concerns. No one could boot the system without the USB drive in place (assuming you have configured your system such as that inserting another USB drive won't work either). However, before you can use USB storage as a main drive solution, your system must provide the following functionality:

- Front panel access to the USB drive to make it easy to insert and remove
- The ability to boot from the USB drive, which means changing the system configuration in the system setup
- Compatibility with the USB drive you choose as a boot device

Working with External Drives

External drives are great because they're much larger than thumb drives and yet they're still portable. You use external drives to move large amounts of data between systems, to provide backup services, and to provide temporary storage for larger tasks. An external drive also makes it possible to improve overall reliability because you can simply move your data to another machine when the original host machine fails.

Most modern external drives work just like thumb drives do. You plug the drive into a USB port on either the front or the back of the system. The operating system automatically recognizes the drive and you can begin accessing it immediately. As with thumb drives, you could conceivably configure a system to boot from an external drive (read the "Using USB Storage" section of this chapter for details). In short, an external drive is just a really big thumb drive. You need to consider some issues when working with an external drive:

- If the external housing supports a standard hard drive, you must close the drive in the operating system before disconnecting it. The act of closing the drive will shut the drive down and move the read/write heads into the parked position.
- Dropping the drive could cause serious damage and data loss.
- An external hard drive is more likely to require larger amounts of power, so you need to make sure the USB port can actually provide the power needed by the drive.
- External drives can get lost, which means that your data could end up in a nonsecure location.

Considering Other Permanent Storage Options

Most systems today rely heavily on SATA connectivity. However, SATA is far from the only drive communication standard available. The main advantage of SATA is that it's inexpensive, which is the main criterion for many people. However, you may find that you want to use an alternative. For example, SCSI is actually a higher quality solution than SATA, while RAID is more reliable. The following sections help you understand the common SATA alternatives.

> **NOTE**
>
> *This book doesn't consider IDE in depth because it's an older technology that you're unlikely to use on a new system. SCSI and RAID are solutions that you are likely to consider for your system.*

Considering the SCSI Option

The SCSI option provides a higher quality interface that is less likely to degrade over time than the SATA option. The main difference is that SCSI setups contain multiple processors to perform various tasks. For example, there is a separate processor to perform commands. A completely separate processor handles the interface with the drive, including head positioning and adjusting for drive wear and tear. As a result, the SCSI interface is robust and reliable. However, those extra processors cost money and consume additional space. Consequently, an SCSI drive tends to be larger than its SATA counterpart is.

> *TIP*
>
> *SCSI drives often come with longer warranties because they are built better. The use of heavier components and multiple processors means that the drive experiences less wear and tear. Therefore, it pays to use SCSI setups where drive reliability is a major concern and RAID isn't an option.*

SATA handles commands one at a time. Yes, there is a bus release command that some people view as command queuing, but in reality, only SCSI provides true command queuing. An SCSI drive can actually organize commands for optimal execution, so that it becomes more efficient than a corresponding SATA drive, even if all other factors (such as spindle speed and head latency) are the same. Organizing the drive commands also reduces wear and tear on the drive, making it last even longer.

Due to the efficiencies that SCSI drives provide and the better quality of the components used, they also tend to be less noisy than a corresponding SATA drive. Obviously, an SCSI drive will never match the quietness of an SSD drive, but can be quieter than SATA, making it a good choice when noise can be problematic.

> *NOTE*
>
> *You normally need a separate host adapter when implementing an SCSI solution. A standard motherboard normally supports only SATA. However, there are higher-end desktop motherboards that do contain both SATA and SCSI host adapters.*

Understanding RAID

A redundant array of inexpensive disks (RAID) is a technology that uses multiple drives to increase speed, reliability, or both. Speed increases when accessing both drives at the same time (effectively doubling the throughput). Reliability increases when one drive mirrors the information contained on another drive. Both SATA and SCSI drives can come in RAID configurations. However, most systems will use SATA RAID today because that's the configuration motherboard vendors support. When comparing a SATA RAID configuration to an SCSI RAID configuration, the same rules apply as to single drives. An SCSI RAID configuration will be more expensive, but also provides higher reliability, better performance, and longer life than a corresponding SATA RAID configuration. The following sections describe RAID in more detail.

Defining the RAID Levels

All levels of RAID rely on multiple drives. However, each RAID level has a different goal in making drives more reliable, faster, or both. The following list describes the RAID levels commonly supported by motherboards today:

- RAID 0: Drives are striped to increase access speed. Striping is the process whereby one bit of data is written to one drive, another bit to another drive, and so on. The information appears across all the drives so that the drive throughput increases by a factor of the number of drives involved. For example, using two drives doubles throughput (or nearly so since there are some losses to coordination efforts between the two drives). The minimum number of drives for this configuration is two.
- RAID 1: Drive sets are paired to increase reliability. The system sends information to one drive and then the same information to the second drive so that the first and second drives always contain the same information. This configuration can actually make the system slower, but with a significant increase in reliability. The minimum number of drives for this configuration is two.
- RAID 5/6: The system stripes information across multiple drives in blocks to improve both speed and reliability. If a single drive is lost, a parity bit allows the system to reconstruct the lost data. The advantage of this configuration is that the parity bits mean more drive space holds actual data, rather than reliability information. RAID 6 is simply an extension of RAID 5 in that it uses two parity bits, which means that two drives can fail without incurring data loss. The minimum number of drives for RAID 5 is three and for RAID 6 is four.
- RAID 0+1: This is a combination of RAID 0 and RAID 1. The drives are first striped to create two striped sets. The striped sets are then mirrored. The effect is to increase throughput and reliability. Depending on the source you read, this configuration provides slightly better throughput at the cost of reliability. Because there are only two striped sets, more than two drive failures will cause the setup to crash. The minimum number of drives for this configuration is four.

- RAID 1+0: This is a combination of RAID 1 and RAID 0. The drives are first mirrored to increase reliability. The mirrored sets are then striped to increase throughput. Depending on the configuration, the reliability of this configuration can be significantly higher than RAID 0+1 because one drive in each mirrored set can fail. As a comparison, if three drives in a RAID 0+1 configuration fail when using six drives, the RAID will crash. However, a RAID 1+0 configuration of six drives will continue to work even if three drives (one in each mirrored set) fail.

Considering the Configuration Requirements

Higher-end motherboards will provide RAID support. The kind of RAID support depends on the vendor and the quality of the motherboard. In most cases, you can expect a motherboard to support RAID levels 0, 1, 5, and 1+0. However, the level of support could vary and you need to consult your manual.

In general, you need to access the system configuration on the system by pressing Delete (or some other access key) during the boot process. The configuration display will let you enable RAID support if your system contains the correct number of drives.

After you enable RAID support, you'll see a new message appear during the boot process to press a special key combination to enter the RAID configuration utility. You must configure the drives for RAID before you do anything else with them. In other words, you can't start with a formatted drive containing and operating system and expect to turn it into a RAID setup.

The RAID setup generally asks you to choose a RAID configuration. It will then display a list of usable drives and ask you to add them to the RAID configuration. After you perform this task, you must still format the drives (which will now appear as a single drive) before you can use the RAID setup. The RAID configuration utility provides everything needed to format the drives for use. The formatting process makes the set of drives appear as a single drive to the system.

9

Attaching Auxiliary Devices

Most computer systems provide access to one or more auxiliary devices that supply specific functionality. Auxiliary devices generally appear outside the case and attach to an external port in some manner. They fall into these categories:

- Devices, such as keyboards, that the system usually requires to perform useful tasks
- Devices, such as printers, that the system can function without, but are critical to user needs
- Nice-to-have appliances, such as webcams, that aren't critical to support user needs, but make the user's environment easier to work with

Precisely where an auxiliary device fits within the usage spectrum isn't material to this chapter except that you should connect critical devices first and worry about nice-to-have appliances as time permits. How you classify a device depends on the user. For example, a developer could possibly get by just fine without a webcam, but a business professional who must interact with others could see the webcam as a critical device.

All of these devices have one thing in common: you connect something to a port on the machine in order to provide a connection between the device and the rest of the system. In some cases, the connection is through a receiver of some sort so that a wireless connection frees the user from having to deal with cables. Even so, the receiver still presents a physical connection to the system—connections don't occur by magic.

Choosing Keyboard and Other Input Devices Carefully

Novice builders can have the misconception that any device can plug into any PC, but that's far from the truth. A PC provides specific ports and those ports support specific devices. In addition, the ports may have limitations. For example, a Universal Serial Bus (USB) port may not be able to provide the amount of power that the device requires to operate. Some devices may also require a specific version of the USB port to function correctly. In short, before you decide on auxiliary devices for your new machine, you need to consider whether the device specifications match those of your system.

Unfortunately, obtaining an auxiliary device that works with your PC isn't always so simple. Specifications provide a written contract of how devices are supposed to work with PCs. A specification controls how the wiring works, the device driver communicates, the protocols are used, and all sorts of endless details that can boggle the mind. The specifications used to create the contract between the auxiliary device and the PC could have holes in it—places where the specification doesn't actually state how things are supposed to work. These gaps in communication force vendors to create solutions based on best practices. When two vendors follow different routes, a device that should work with your system may not do so because the device and the system are incompatible.

WARNING

Some device compatibility problems can become severe. For example, a misbehaving thumb drive could cause your system to freeze, rather than complete the boot process. In some cases, compatibility issues can actually cause system damage, although such damage is incredibly rare. Always add devices one at a time and verify the device works before you move onto the next device.

It's also entirely possible for a device to work with one operating system and not another. The device works, your system recognizes it, but the operating system can't communicate with the device for any of a number of reasons. A device and its associated driver disk will often tell you which operating systems that the device can support. If you find that the device isn't working with the operating system, it could be a driver problem and you need to look for an updated driver online.

Performing the research required to determine whether a device you want to use will actually work saves time and money. It may seem in the short term that you're actually using up a lot of time researching your purchase, but troubleshooting errant devices is both time consuming and error prone. Getting a device that works properly the first time you plug it in will always save time and make using your computer a pleasure.

Connecting the Keyboard

Your system requires a keyboard in order to enter commands. Even when your system can't boot, you need to be able to enter commands to configure the system. In general, if you don't have a keyboard attached to your system, the system will recognize the fact and refuse to continue the startup procedure. The following sections discuss a variety of keyboard options and concerns.

Considering the Advantages of Wired Keyboards

Many people use wireless keyboards today for good reason—you can place the keyboard anywhere it feels comfortable to work, even your lap. A wireless keyboard is convenient when you need flexibility in your work environment. It might seem as if the only good keyboard option is a wireless model simply due to the flexibility and convenience it provides, but a wired keyboard has some significant advantages you need to consider:

- System configuration: It turns out that the system may not recognize a wireless keyboard early enough in the boot process to enable it for use in the system configuration process. As a result, even if you rely on a wireless keyboard normally, you may require a wired keyboard on hand to perform the infrequent (but required) system configuration tasks.
- Reliability: A wired keyboard has a continuous connection to the system. Unless you cut the cable or disconnect the connector, a wired keyboard is extremely reliable. Other devices can interfere with wireless keyboard signals and the use of batteries means that the wireless connection will eventually fail when the batteries are dead.
- Ergonomics: Many wired keyboards come in configurations that make typing easier and less stressful. The keys used with some wired keyboards are also easier on your hands.
- Features: Wired keyboards can include a wealth of features not found on wireless keyboards because a wired keyboard can have better communication with the system. Just what you obtain in the way of added functionality depends on the kind of keyboard you get.
- No electromagnetic interference (EMI) issues: Wireless keyboards are subject to EMI signal interference. There are many potential sources for EMI and tracking down the precise source of your problem can be difficult. The EMI could cause the keyboard to fail intermittently or could cause complete failure.

Good Keyboards Cost More

If you really want a good keyboard, you need to pay for it. Many people fail to look at the whole picture when it comes to keyboards. They end up paying the price with a keyboard that fails to work as advertised and physical injury from repetitive stress. Here are some things to consider when buying a keyboard:

- Shape: The shape of the keyboard determines the angle at which your hands contact the keys. A split or curved keyboard provides better support for your hands and reduces wrist injuries.
- Key type: Keyboards have different key types. What works best for you can depend on personal preference. However, keyboards that use spring-loaded keys tend to cause less stress on your body and are actually faster to use.
- Wrist support: Most keyboards today come with a wrist support. However, when the wrist support is too wide (or narrow) or at the wrong angle, it can actually do more harm than good.
- Key placement: Depending on the sort of work you do, the placement of the keys can make a great impact on how easy the keyboard is to use and the stress on body parts as you use it.
- Features: Some keyboards have special features that make them easier to use for some tasks. For example, an Internet keyboard can make it easier to perform research or other tasks on the Internet.
- Software support: The software support for a keyboard varies greatly. For example, buying a programmable keyboard can save you time and effort by making it possible to create keyboard macros that attach to the keyboard, rather than a specific application. However, the software support is only useful when it supports the operating system you're using.

Working with Wired Keyboards

Wired keyboards connect directly to the system using a special plug as shown in Figure 9-1. A system will usually have two connectors that look the same sitting right next to each other as shown in the figure. One connector will have a picture of a mouse next to and the other connector will have a picture of a keyboard. Be sure you plug your wired keyboard into the keyboard connector.

> **NOTE**
>
> Keyboard and mouse support varies by system. You might actually find that the keyboard and mouse work just fine no matter which of the two plugs you use. However, the safest bet is to ensure you plug the keyboard into the keyboard connector. If you find that your keyboard doesn't work, then try plugging it into the other connector.

Chapter 9: Attaching Auxiliary Devices

FIGURE 9-1 Use the special keyboard plug to attach your keyboard to the system.

Working with Wireless Keyboards

A wireless keyboard doesn't have a physical connection to the system. Instead, the keyboard communicates with the system using a receiver. The receiver plugs into the system instead of the keyboard. Figure 9-2 shows an example of a wireless keyboard and receiver. In this case, you're seeing a keyboard and a Bluetooth adapter for a USB port.

FIGURE 9-2 A wireless keyboard consists of a keyboard and receiver combination.

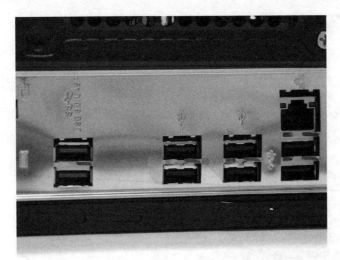

FIGURE 9-3 Rely on the back of system USB connectors when connecting the keyboard or mouse.

The receiver generally plugs into a USB or Bluetooth connector. The most common connection is the USB connector. This connector must support the power requirements for the receiver. Otherwise, you could damage the receiver, the system, or both when you plug the receiver into the connector.

When working with a USB setup, you generally use one of the USB connectors on the back of the system as shown in Figure 9-3. This allows you to reserve the front connectors for devices that you plan to plug and unplug on a regular basis. The keyboard will likely remain plugged in all of the time, so there is no reason to waste a front panel connector on it.

Most motherboards don't come with Bluetooth support. Obtaining the required support is a matter of adding a USB adapter (http://www.amazon.com/exec/obidos/ASIN/B007Q45EF4/datacservip0f-20/). All you need to do is plug the adapter in.

> **WARNING**
>
> *A USB-connected Bluetooth adapter may not become available until the operating system is online, which could cause problems with a keyboard, but not with a mouse. The system requires a keyboard to boot unless you specifically configure the system setup (using a keyboard) to allow the system to boot without one. Make sure you know how the Bluetooth support for your system works before you rely on a Bluetooth-connected keyboard.*

Connecting the Mouse or Trackball

From a system configuration perspective, a mouse or trackball isn't necessary. In fact, when you first turn the system on, the mouse or trackball might not even be

Chapter 9: Attaching Auxiliary Devices

active. The mouse or trackball may not work within the system configuration screens either. However, once you boot the system, the operating system you use may depend heavily on the mouse or trackball. It depends on the operating system you're using.

> **NOTE**
>
> *Contrary to popular opinion, it's possible to work with operating systems like Windows without using a mouse. You use keyboard shortcuts to access features you normally access using the mouse or trackball. However, working with graphical operating systems using the keyboard can be slow and cumbersome. In addition, any instructions you have to working with applications will likely rely on using the mouse or trackball.*

Mouse Alternatives

The trackball is one of the most common mouse alternatives, which is why it is covered in this chapter. However, there are many other mouse alternatives out there and the input device you use depends on the sort of work that you perform. For example, some people prefer using a touchpad because it's an easier way to provide input when not sitting at a desk (it's also the more common input method for laptop systems).

One of the most interesting alternatives is the rollermouse (http://www.amazon.com/exec/obidos/ASIN/B003AU0FLQ/datacservip0f-20/). It's an especially popular alternative for people with repetitive stress injuries (RSIs) because it doesn't require any sort of wrist movement to use. To use a rollermouse, you simply roll the roller forward or backward or move it side-to-side.

If you're a graphic artist or work with computer-aided design (CAD) software, you may prefer a pen tablet (http://www.amazon.com/exec/obidos/ASIN/B00EN27U9U/datacservip0f-20/) instead. In this case, you actually draw what you want to see on screen using pen and paper. When working with CAD-specific input devices, you can sometimes select items and drag them over to the drawing area.

The point is that you can choose mouse alternatives that make you more productive when performing specific tasks. These devices attach to your system using the same techniques as used for a mouse or a trackball. The computer interface is the same, but the human interaction is different.

Considering the Trackball Difference

A trackball (http://www.amazon.com/exec/obidos/ASIN/B00009KH63/datacservip0f-20/) is essentially an upside-down mouse. At the top of the trackball is a ball that you roll around using your fingers to position the mouse cursor on screen. Because of the trackball design, you don't need much desk space to use one. A trackball also tends to be easier for people with RSI to use.

Depending on how you work, a trackball can also provide greater accuracy in positioning the mouse cursor.

The main downside of using a trackball is that the ball gets dirty after a while, making it harder to position the cursor. The dirt gets inside the device, eventually causing it to fail. Yes, you can clean your trackball, but it takes a lot of discipline to do so. In addition, trackballs are more susceptible to spills (although spilling anything on any computer device is a bad idea).

Working with a Wired Mouse or Trackball

Unlike your keyboard, you don't absolutely need a wired mouse or trackball to work on the system configuration screen. Most system setup software won't even recognize the mouse or trackball. A wired mouse or trackball will normally connect to the mouse connector on the back of the system shown in Figure 9-1.

However, there are two other ways to connect a wired mouse or trackball. The first is to use a USB port. You simply connect the mouse or trackball to the back of the system using a USB port like the one shown in Figure 9-3.

The second alternative mouse and trackball connection is the COM port. This is a nine-pin connector on the back of the system like the one shown in Figure 9-4. For the most part, you only find this connection type supplied with older mice and trackballs. However, it's important to know that the option exists should you need it. The main disadvantages of using the COM connector are that the system won't automatically recognize your mouse, you may need to configure the system to activate the COM port, and you can't attach/detach the mouse or trackball while the system is running.

> **TIP**
>
> *The resolution of your mouse or trackball determines how sensitive the mouse or trackball is to movement and how much effort it takes to move the cursor across the display. Generally, a higher resolution mouse or trackball is better than a lower resolution device, but you may need to try out several models to determine which one works best for your needs. If someone who is left-handed uses your mouse or trackball, getting an ambidextrous model is usually the best idea so that right-handed helpers can use the device as well.*

FIGURE 9-4   Some older mice and trackballs use the COM port connection.

Chapter 9: Attaching Auxiliary Devices

FIGURE 9-5   Mice and keyboards can use custom receivers that connect to the USB port.

Working with a Wireless Mouse or Trackball

A wireless mouse or trackball works much the same as a wireless keyboard. You plug a wireless receiver into a USB or Bluetooth connector. The receiver provides wireless communication with the mouse or trackball. A wireless mouse or trackball setup has the same advantages and disadvantages as a wireless keyboard setup. The most common problem that you encounter is EMI signal interfering with the transmission between device and receiver. Figure 9-5 shows a wireless receiver and a mouse. Notice that this is a USB-connected receiver specifically designed for the mouse in this case.

Choosing Between Generic and Specific Wireless Receivers

This chapter shows two kinds of wireless receivers. The wireless keyboard uses a Bluetooth receiver, which is generic and can support other devices. The wireless mouse uses a special-purpose receiver that only supports the mouse and no other device. Both receiver types are useful in particular situations.

A generic receiver is great when you want to support more than one wireless device using it. For example, the Bluetooth receiver will detect and support other Bluetooth devices. A single adapter can support as many devices as are needed, so you don't have to fill your workspace with countless receivers that could interfere with each other. However, the down side to using Bluetooth is that there is only so much bandwidth and you could find that devices hog the bandwidth (slowing other devices down) or interfere with other devices' usage of the connection.

A special-purpose receiver ensures that the supported device has all the bandwidth it requires and the receiver could potentially reduce interference with other devices.

However, you must choose special-purpose receivers carefully to ensure they don't operate at the same frequency and don't interfere with noncomputer devices in the immediate area. Using a number of special-purpose receivers can also cause clutter around your workspace and use up the available system ports.

Use a generic receiver when the need for bandwidth by a particular device is small and you don't have more than a few devices that could possibly need to share the port. A special-purpose receiver is the best choice when you must assure the device has a consistent, high-quality connection to the system.

Testing the Basic Setup

After you add your keyboard and mouse, you should be able to perform a nearly full boot of your system. The following procedure helps you test your system to ensure it is ready to accept additional auxiliary devices, a network connection, and ultimately, an operating system.

1. Verify that the keyboard is connected to the appropriate connector on the system. If you're using a wireless keyboard, ensure the receiver is connected to the appropriate connector on the system.
2. Verify that the mouse is connected to the appropriate connector on the system. If you're using a wireless mouse, ensure the receiver is connected to the appropriate connector on the system.
3. Plug in the computer and apply power to the system. Verify that the various computer fans have started (there should be fans for the power supply, display adapter, and the processor as a minimum, but your system could contain additional fans).
4. Allow the power to cycle long enough so the display adapter becomes active and shows information on the display. You should see the power on startup test (POST) complete. You should see various kinds of information, including the following entries:
   - The types of storage devices installed
   - Any ports you have enabled
   - The amount of memory installed
   - The kind of processor installed
   - Additional configuration information for your particular motherboard
5. If you see a keyboard error message, repeat steps 1 through 4 until the system passes POST.

When using a wireless keyboard, you may have to follow these additional steps to pass POST:

1. Turn the system off.
2. Disconnect the wireless keyboard.

Chapter 9: Attaching Auxiliary Devices

3. Plug in a wired keyboard.
4. Turn the system on.
5. Enter the system configuration by pressing Delete (or the special key for your motherboard) during the boot process.
6. Configure the system not to check for a keyboard during boot.
7. Save the new configuration.
8. Turn the system off.
9. Unplug the wired keyboard.
10. Connect the wireless keyboard.
11. Turn the system on.

NOTE

The system will pass POST, but it won't boot completely because you haven't installed an operating system yet. You should see an error message stating that the system can't find a boot device. Even though the hard drive will eventually contain an operating system, you don't have one installed right now, so the best your system can do is pass POST.

Adding a Printer

Many printers today are designed to work stand-alone on a network. You plug them into the network switch and every system on the network can see the printer automatically. However, you may choose to create a stand-alone system, which means you need a printer that connects directly to your system.

Connected printers commonly rely on a USB connection. All you need to do is plug them into a USB port on the back of your system (see Figure 9-3) and the system will automatically recognize the printer. Unlike the mouse and keyboard, your operating system may not provide a generic driver for a printer or possibly lack one for your specific printer. Make sure you have a device driver disk ready to install printer support.

Older printers may rely on a parallel port like the one shown in Figure 9-6. Using a parallel port is exceptionally rare today, but you can still find printers that use one. You must turn the system off, connect the printer, and turn the system back on to use a parallel port. The system may not automatically recognize your printer and you definitely need a device driver to make the printer active.

It's extremely unlikely that your motherboard will provide support for a parallel connector. If you have a printer that requires a parallel connection, you need an add-on card like the one shown at http://www.amazon.com/exec/obidos/ASIN/B004B0H69I/datacservip0f-20/. You add this card into your system by plugging it into a PCI connector and then screwing the card down, much as you did for the display adapter.

FIGURE 9-6 Some older printers still use a parallel port, but such setups are becoming uncommon.

Working with Webcams

Many people have webcams connected to their systems today. Webcams can connect directly to a USB port on your system or you can obtain a wireless model where the webcam connects to the system through a receiver. There are also Bluetooth configurations, assuming you have a Bluetooth receiver.

The two essential considerations for a webcam from the perspective of building your own PC is whether the webcam supports a connection that your PC supports and whether the vendor designed the webcam to work with your operating system. Generally, you need the correct device drivers to make the webcam active.

From a functional perspective, you need to ensure the webcam will connect to a monitor, desktop, or other surface that you have available. It's almost important to determine whether the webcam comes with a microphone when speech is part of the configuration. The size of the picture that the webcam produces is also important. Getting a 1080p configuration will produce a better picture than a 720p setup. However, the 1080p configuration also requires more bandwidth when working with applications such as Skype and you need to ensure that your connection will support it.

Providing Other Device Connections

You can connect myriad devices to PCs either directly or wirelessly. There are some general rules to follow as the number of auxiliary devices increases:

- The more wireless devices you use, the greater the chance of EMI problems. Connect essential devices first and then devices of lesser importance. Always connect devices one at a time so that you know precisely which device is causing a problem.
- Connection support is an essential element of supporting auxiliary devices. You can't use a Bluetooth device if you don't have a Bluetooth receiver.

Chapter 9: Attaching Auxiliary Devices

- Most devices require driver support. This means that the device documentation should say it supports your operating system directly. Otherwise, you can't be sure that the operating system will even see the device.
- Using USB or SATA connections for devices will reduce the amount of work required to install and configure the device. Use older connection types only when required to meet specific needs (such as support for older scientific equipment).
- Use wired connections when you must ensure the device will be available during the boot process to perform configuration-specific tasks, such as working at the configuration screen. If you decide to use a wireless keyboard, for example, make sure you have a wired keyboard as backup so that you can perform required configuration tasks.
- Make sure your device can obtain required physical support. For example, if you don't have room for another drive in your system because all the drive bays are in use or the motherboard lacks additional connections, you can't add that particular device.

Part III
Considering Networks

10

Installing a LAN

At one time, almost every computer was stand-alone, which meant that it didn't connect to any other computer. You can still create stand-alone computer setups, but they have become rarer. Most computers today connect to other computers in some way. At one time, you could encounter all sorts of connection types. However, today the connections generally connect locally, using a local area network (LAN), or globally through the Internet, using a special device that somehow attaches to the LAN. In both cases, the computer relies on LAN hardware to make the connection. This chapter helps you discover the techniques and requirements for creating computer connectivity with your new PC.

> **NOTE**
>
> *Even though this chapter relies heavily on Windows 7 screenshots, the same principles apply to nearly any network you want to install on your system. Networks require the use of special software to communicate between machines. This software normally lies under the surface where you can't see it and works automatically. However, it's important to know that the software exists and that you may need to install it separately with some operating systems.*

Understanding LAN Basics

A problem exists today in that LANs are almost too easy to connect after someone configures them. With a wireless local area network (WLAN) based on the 802.11 standards, you can literally walk into a room with the proper setup and gain an instant connection. A WLAN requires no wires to create the required connectivity. However, it's still important to understand what happens when you create a LAN or WLAN setup because you need to perform the required configuration and be able to troubleshoot the setup when it fails.

Considering the LAN Hardware

The LAN hardware makes it possible for devices to connect to each other. Older PCs didn't have a LAN adapter installed; you installed it as a separate item. Modern PCs usually have one or more LAN adapters provided as part of the motherboard as shown in Figure 10-1. However, you may require more than one port, which means buying a separate adapter.

An adapter, such as the one shown at http://www.amazon.com/exec/obidos/ASIN/B0000TO0BQ/datacservip0f-20/, provides you with a single network connection. Adapters also come with USB connectivity, as shown at http://www.amazon.com/exec/obidos/ASIN/B00NOP70EC/datacservip0f-20/. The adapter form has the advantage providing a back-of-the-case connection that isn't easy to disconnect from the system. You can also obtain adapters that have dual ports, such as the one shown at http://www.amazon.com/exec/obidos/ASIN/B00DODX5MA/datacservip0f-20/. This adapter provides you with two network connections. There are cards that can provide up to four ports, should you need that many.

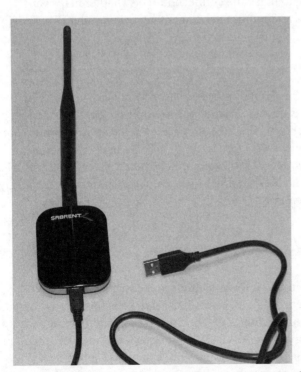

FIGURE 10-1 Modern PCs come with one or two LAN adapters in most cases.

> **WARNING**
>
> *If you purchase a LAN adapter that resides on a separate card, the card must be compatible with the expansion slots on your system. Most modern systems will require that you buy a Peripheral Component Interconnect (PCI) compatible card. However, you need to check the documentation for your motherboard to ensure you meet any requirements for it. In addition, make sure you actually have a free slot. Some motherboards provide only a few free slots because motherboards provide so many built-in features.*

You must have one port for each network you want to access. If you have a home network, that would require one port to access. In order to connect directly to the Internet through a device such as a modem from your system, you need a second port. Your home network counts as one network and the Internet counts as a second network. A system that has two network connections—one to the home network and another to the Internet—can act as a router. A router acts like a traffic cop, sending signals from the home network to the Internet and signals from the Internet to the home network.

Only one system on your network has to provide router services, which means that only one system has to have two ports if you have just a home network and a connection to the Internet as well. It's also possible to buy stand-alone routers, such as the one shown at http://www.amazon.com/exec/obidos/ASIN/B002YLAUU8/datacservip0f-20/. In this case, the router provides both wired (gigabit) and wireless connectivity.

Creating the Physical Connection

To create a connection between two machines, each machine has to have a network adapter that accommodates that connection. Between the two machines is a hub (a passive device that simply makes connections) or a switch (an active device that includes intelligence for performing tasks such as diagnostics). Remember that routers are used to create connections between two networks.

The machines use either wires or radio waves to create the connection. When a machine uses wires to make a connection, the cable has to adhere to the requirements for making the connection at the requested speed. This means using Cat 5, 5e, or 6 cabling. The Cat 5e cabling is acceptable if you want to create a gigabit (1000-Mb/s) network. However, to get the best response from a gigabit network, use Cat 6 cabling instead. Even though you can use Cat 5 cabling for networks, it doesn't respond well to high speeds and you won't see the full throughput your network is able to provide.

When a machine uses radio waves to create a connection, it must use a signal that provides the appropriate functionality and strength for the setup. The "Understanding WLAN Essentials" section of this chapter provides additional insights into wireless connectivity.

Considering the LAN Software

LAN adapters rely on a device driver, just like any other device, to create a connection to the operating system. However, you need more than just a device driver to make the connection. A LAN adapter also requires software that can create and interpret packets of network data, send and receive packets, and perform tasks such as negotiate communication with other machines on the network. Figure 10-2 shows a somewhat typical setup.

The software falls into a number of categories depending on its use. Not shown in Figure 10-2 is the device driver, but there is one involved to provide the low-level access between the physical device and the operating system. The software you can see falls into three categories:

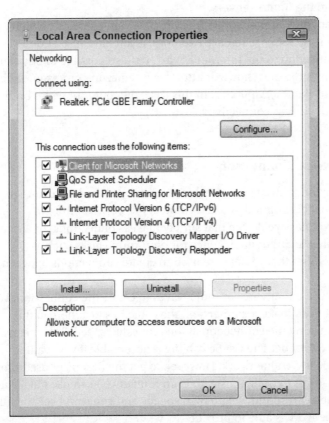

FIGURE 10-2 A LAN relies on software to create connections.

- Client: A client is an application that provides high-level network access. Clients can let the system act as a workstation (something that uses resources) or a server (something that provides resources). Other kinds of clients let the system perform other high-level tasks.
- Service: A service runs in the background and lets the system perform specific tasks without the user's knowledge, such as sharing printers and files. Services can also enhance the functionality of the network and make it more responsive.
- Protocol: A protocol is a set of rules that determines how the network performs tasks. Committees create these rules to ensure that various operating systems can talk to each other. The most common protocols in use today are Transmission Control Protocol/Internet Protocol (TCP/IP) version 4 and TCP/IP version 6.

Seeing the Other Connections

The purpose of a network is to let you see other connections so that you can provide or access resources, such as printers and files. Even a web page is a series of files that you access. A browser downloads one or more files, interprets their content, and then displays the content on screen to you. However, networks provide access to all sorts of files, data streams, and other resources that you can use.

Even when you have a physical connection to the network and have the appropriate software installed, you may still not see other connections. In order to see the other connections, your computer must follow the rules set out by the protocols that you're using. For example, you need to have an appropriate network address. Fortunately, most operating systems today perform all the required setup for you. Even so, it's good to know a few things about your connection so that you can troubleshoot it when the time comes. You can find some useful TCP/IP tutorials at:

- http://www.w3schools.com/website/web_tcpip.asp
- http://www.garykessler.net/library/tcpip.html
- http://www.hardwaresecrets.com/article/433

Some operating systems, such as Windows, also provide automatic methods of discovering network problems. Figure 10-3 shows the Network and Sharing Center where you can rely on automation to set up and configure networks, diagnose errors, and even map your network so you can see other computers that share it. These diagnostic aids can locate many common problems and even a few difficult problems. However, if you don't understand what the diagnostic is telling you, then the automation isn't all that helpful. Spending some time reviewing how your connectivity works, especially when that connectivity is operational, is a good way to prevent problems later.

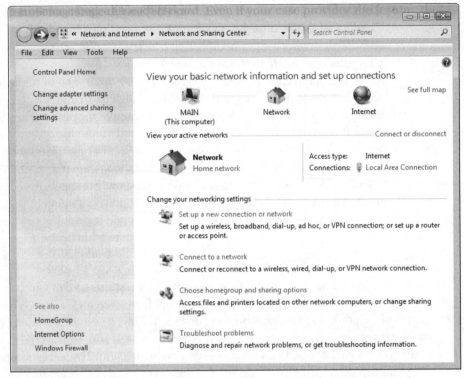

FIGURE 10-3 Newer operating systems provide automation that makes network management easier.

Configuring the Motherboard LAN

Depending on the motherboard vendor, you may not actually have to perform any LAN adapter configuration. However, it pays to look at the configuration features before you complete your system setup so that you know what options you have in case problems arise. In some cases, the motherboard manual will also contain information about configuring the LAN adapter to ensure it works properly with your operating system. The following sections provide some guidelines on configuring the LAN adapter on your motherboard.

Dealing with Jumpers

Motherboard vendors have tried to eliminate jumpers by creating better system setup utilities. However, sometimes you still find jumpers on the motherboard. The only way you know how to adjust the jumpers is to check your manual. In most cases, the LAN adapter has settings that control these features:

- Whether the LAN adapter is enabled
- Interrupt request (IRQ)

- I/O address
- Memory address
- Special features (such as Wake-on-LAN)

Performing the System Configuration

The system configuration display that you can access by pressing a special key, such as F2 or Delete, during the boot cycle contains LAN adapter settings on most modern motherboards that have a LAN adapter. This is especially true when your motherboard lacks any jumpers—the functionality provided by the jumpers has moved to the system configuration display.

In most cases, you want to perform any LAN adapter configuration before you install the operating system. By making these changes first, the operating system can detect the settings and add your network support correctly when the time comes.

> **NOTE**
>
> *Many of the newest jumperless systems rely on a negotiation phase with the operating system to configure options such as IRQ, I/O address, and memory address. The advantage of this setup is that newer operating systems automatically configure all components it can for optimal performance—working around any components that offer less flexible configuration options. The disadvantage is that you might find it difficult to work with older operating systems that lack the negotiation phase. In this case, you normally rely on the default LAN adapter configuration settings or disable the LAN adapter and use a separate LAN card instead to provide networking support.*

Considering Multiple-LAN Motherboards

Any system having a motherboard that supports more than one LAN adapter can act as a router as long as the operating system you install provides the required functionality. The main reason to buy a motherboard with multiple LAN adapters is to provide this kind of functionality (and avoid having to buy a router for your network).

> **TIP**
>
> *In some cases, you may not need more than just one LAN adapter. In this case, you want to disable the LAN adapters you don't need in the motherboard setup, rather than in the operating system you install. If you try to disable the LAN adapter in the operating system, the operating system could spend a lot of time telling you that the LAN adapter is disabled and want to enable it for you. The disabled LAN adapter can also cause problems when you try to diagnose connectivity issues.*

FIGURE 10-4 When working with two LAN adapters, mark the LAN adapter uses on the back of the case.

It's important to realize, however, that many motherboard that include multiple LAN adapters provide LAN adapters of different types from different vendors. The LAN adapters may operate at different speeds and provide a different feature set. For example, one of the LAN adapters may operate at a higher speed than the other or provide a feature such as Wake-On-LAN (WOL). (You can read more about WOL at http://www.howtogeek.com/70374/how-to-geek-explains-what-is-wake-on-lan-and-how-do-i-enable-it/.) Choose the LAN adapter you use for a specific purpose carefully. The higher-speed LAN adapter should always connect to your internal LAN because an outside connection (e.g., to the Internet) is bound to be slower.

In general, when a motherboard has two LAN adapters, both LAN adapters have rear panel access as shown in Figure 10-4. You need to ascertain which LAN adapter connects to each of the plugs because you can't change this arrangement. The motherboard manual tells you about the LAN adapter outputs, but performing a test usually works best to ensure that you know which plug to use for a given purpose. Marking the plug, so that you can later determine how to reconnect network connectors, is a great idea.

A motherboard may also put the LAN adapter outputs on a standoff. When this is the case, you usually find an appropriate ribbon cable with the required plug included as part of the motherboard packaging. The plug will connect to the standoff and you place it in a special holder in an expansion slot opening. Make sure you choose an expansion slot that you won't need for other purposes to avoid having to move things around later.

Installing a Separate LAN Card

If your motherboard doesn't provide enough LAN adapters, then you need to install a separate LAN card to provide one or more additional connections. The "Considering the LAN Hardware" section of this chapter discusses some of the approaches you can use to obtain additional connections. Once you obtain a separate LAN card, you need to perform these steps to ensure it works properly:

1. Configure any required jumpers on the card (modern cards normally don't rely on jumpers, but you should check the manual anyway, just to be certain).
2. Open the case and remove an expansion slot cover for the expansion slot that will hold the LAN card.
3. Carefully seat the card into the slot, just as you would any other expansion card.
4. Screw the back of the card into the case.
5. Connect the LAN cable to the back of the card.

The LAN adapter on the card may not be functional until you install an operating system (see Chapter 13), install the special LAN adapter device driver (see Chapter 14), and perform any required setup (see the vendor manual). If you have more than one LAN adapter on the card, then you may need to configure each LAN adapter individually.

> *NOTE*
>
> *You can't configure a separate LAN card using the system configuration utility. The system configuration utility only affects LAN adapters that appear as part of the motherboard. If your LAN card supports a boot up configuration utility, you access that utility using a different key press, such as F5, during the boot process. The LAN card configuration screen looks different from the system configuration screen and you follow the instructions in the LAN card manual to use it.*

Using External LAN Solutions

To create a connection to another computer, your computer must always have a physical connection to a LAN solution of some type. Even if the adapter is a wireless model, there is still a connection between the system and the adapter. However, you don't always have to have a dedicated connection. For example, in the "Considering the LAN Hardware" section of this chapter you read about an adapter that can connect to your USB port. The USB port acts as an intermediary between the system and the LAN adapter. However, the fact remains that there is still a string of physical connections involved. The most common alternatives for external LAN solutions are (in order of commonality):

- USB
- Serial port (RS232, RS422, and RS485)

- Parallel port
- Small Computer System Interface (SCSI)
- Analog (generally used only for medical or scientific devices)

All of these solutions help you create a network connection by using an external LAN adapter. The connections are both reliable and flexible. However, you must be willing to pay the price for using alternative solutions. Here are some of the penalties you could be asked to pay when using an external LAN solution:

- Reduced throughput
- Shorter transmission distances
- Higher costs
- Increased software installation complexity
- Reduced operating system support

Of the external LAN solutions, using a USB port is the least likely to cause problems, so you should strive to use USB whenever possible. USB to Ethernet connections are readily available, relatively inexpensive, and well-supported by most major operating systems. If you use a USB 3.0 or higher port, you may not even see a significant throughput cost. Of course, you still have to deal with the problems of working with a device hanging off the side of your system.
A passerby could interrupt the connection at the most inopportune times.

Understanding WLAN Essentials

Wireless local area networks (WLANs) are extremely popular today because they provide connectivity in locations where you might not otherwise be able to provide connectivity conveniently. Chapter 12 provides a detailed look at wireless connectivity because it's such an important topic. However, the following sections give you a start on the topic from a hardware perspective and help you understand the alphabet soup of standards from an overview perspective (enough to purchase any WLAN hardware you might want to get).

Defining the WLAN

A WLAN usually refers to a setup where you use wireless, rather than wired, connections between devices on a network. Radio waves stand in for the wires. In most cases, a WLAN is used in a localized setting, such as a home, business, or school. However, in some cases, a WLAN can cover a much larger area, such as a city. WLAN setups in a city require all sorts of expensive equipment that you won't find discussed in this book, such as repeaters. For the purposes of this book, a WLAN covers a localized area.

Radio waves create the connections in a WLAN, but not just any radio waves. The radio waves need to carry data in a manner that both the sender and the receiver recognize as being correct for the data in question. With this need in mind, WLANs usually vary in these specifics:

- Frequency used (2.4 GHz and 5 GHz are most common)
- Transmission speed
- Protocols (see Chapter 12)
- Services
- Applications
- Distances served
- Antenna type

Using a WLAN is usually quite simple because the hardware and software negotiate all the particulars required to create a connection. However, knowing about these particulars can help you overcome problems. For example, some older buildings use tin roofs or tin ceiling decoration that can interfere with wireless signals (or sometimes absorb them). Some wireless connections also require good line-of-sight connectivity to work well.

> **NOTE**
>
> *You have other sorts of wireless connectivity options that don't involve radio waves. For example, it's quite possible to create a connection using infrared light. However, to make this sort of connection work, you need a line-of-sight connection and the distances that such connections work are limited. Some networks also rely on lasers and other sorts of alternative wireless connectivity. None of these other options appear in this book because they present technical challenges that most people can't easily overcome, they're expensive, and they don't work as well as the more popular solutions.*

Delving into Wi-Fi

This book uses the term WLAN throughout to refer to a wireless LAN that meets the Institute of Electrical and Electronics Engineers (IEEE) 802 standards (see Chapter 12 for a discussion of various wireless standards). However, you may also hear the term wireless fidelity (Wi-Fi). The term Wi-Fi is a subset of WLAN and people use the two terms synonymously at times. However, Wi-Fi is a trademarked term that refers to products that have passed the Wi-Fi Alliance (http://www.wi-fi.org/) certification tests. When a product actually does pass this certification, it bears the Wi-Fi Alliance logo.

Unfortunately, you may find that some Wi-Fi products don't have the logo and are therefore not true Wi-Fi products. More importantly, you can't trust the products that are missing the logo to meet Wi-Fi connectivity standards, which means you might get stuck with a less than usable WLAN configuration. You can always verify that products are Wi-Fi certified using the Wi-Fi Allowance Product Finder (http://www.wi-fi.org/product-finder). In short, make sure you exercise caution when obtaining products for your WLAN.

From a system builder perspective, installing Wi-Fi support can be as easy as getting an adapter, such as the one shown at http://www.amazon.com/exec/obidos/ASIN/B00JDVRCI0/datacservip0f-20/, to connect to a USB port. You generally use a USB port in the back of the system because it's unlikely that you'll unplug the adapter. Getting an adapter with an external antenna will extend the device's range and reduce the potential for interference. Figure 10-5 shows a typical Wi-Fi installation that relies on a USB port connection.

It's essential to understand that WLAN and Wi-Fi aren't actually synonymous. A WLAN can adhere to standards other than IEEE 802.11, but Wi-Fi always uses the 802.11 standards. You can find uncertified WLAN devices, but they can work just as well as Wi-Fi devices do. WLAN is an inclusive term that can include a broad array of other device types.

Delving into WiMAX

Worldwide Interoperability for Microwave Access (WiMAX) devices are based on IEEE 802.16 standards and bear the WiMAX logo (just as Wi-Fi devices do). This standard is another form of WLAN, but you see it used mainly for broadband mobile devices. It features greater connection distances and some functionality not found in Wi-Fi. From a system builder perspective, WiMAX requires the use of an expansion card such as the one found at http://www.amazon.com/exec/obidos/ASIN/B005EDCVC8/datacservip0f-20/. It doesn't pay to install WiMAX support in your system unless there are other WiMAX devices to which you can connect. The WiMAX Forum (http://www.wimaxforum.org/) provides full information on WiMAX devices and a listing of WiMAX certified devices.

FIGURE 10-5   A Wi-Fi adapter connected to a USB port in the back of a system.

Considering LAN Security

The most secure system in the world is one that doesn't run applications, has no connections, and doesn't allow anyone to touch it. Of course, a system with this level of security is completely useless and no one would want to own it. The least secure system in the world is one that is open to everyone, allows anyone to touch it, and connects to everything. Unfortunately, this system is also completely useless and no one would want to own it. A useful system is one that has security somewhere between so tight that it prevents all access and so lax that everyone can access it. Most people don't view security in this light. They see security as a bunch of rules designed to take the fun out of computing. However, real security is quite the contrary—it puts the fun into computing by making it possible to do amazing things with your computer.

It's impossible to provide a complete guide to security in a single section of a book devoted to building a computer. The following sections provide you with some solid guidelines to follow as you get started with your new computer. The goal is to help you create a secure environment that makes it possible for you to set up and configure your system, install some applications, and begin having some fun. However, you need to go beyond the simple guidelines provided in this chapter and implement additional security as your system connectivity grows, along with the functionality you build into your system.

Developing a Security Plan

Security is about risk. You have to balance the risks of doing something against the gains that you expect from doing it. When you install a browser on your system, you risk that someone will add a virus or other software to an online site and that you'll download the virus or other software to your machine. Once the virus is downloaded, the perpetrator gains access to all your personal information. Perhaps the perpetrator goes on to erase your data or do something else nasty. This is a worst-case scenario, but you must consider it because it happens every day. Against this worst-case scenario, you balance the need to obtain information, interact with other people, perform downloads, and have fun. A security plan helps you define the line between too much risk and not enough fun. By securing your browser (possibly keeping some sites from working) and not visiting places that typically have viruses on them, you reduce your risk without reducing your fun (at least, not by much). It's a matter of balance between risk and fun—something you need to think about and potentially write down.

LAN security involves localized risk. When you connect to another system, you can access to resources, but you can also infect that machine (or that machine can infect you). Your security plan should also spell out the specifics of localized access. For example, you need to consider which of your resources to make available through the LAN connection. It's also important to consider those seemingly mundane tasks, such as changing your password (even on your home network).

Something that many security books don't discuss is that LAN adapters often come with special features that improve security. For example, you may find that the LAN adapter provides a hardware-based firewall that can be quite useful in keeping perpetrators at bay. Perhaps you can configure the LAN adapter with additional security or use built-in features to help you detect unwanted intrusions. Unfortunately, you won't know about these features or discover how to configure them unless you read the vendor manual. Part of your security plan should include doing the research needed to make maximum use of the functionality that your hardware provides to protect your system from harm.

> **WARNING**
>
> *Some people assume that the security features provided by hardware are turned on by default because the hardware does things like automatically configure itself. The hardware vendor can't assume anything about your setup and therefore turns all security features off. This means that your system starts out wide open and you need to perform the tasks required to secure it from external access.*

LAN adapters do provide security at a connectivity level, but you must also consider the security provided by other hardware, such as your motherboard. For example, many people never configure the administrator password for their system, which means that much of the motherboard functionality is open to anyone who wants to tweak it. The motherboard also provides various protective features. For example, the motherboard can thwart unauthorized access of protected areas of memory.

In some cases, you find other forms of hardware protection provided with your system. It pays to read about all of them and use those measures that make the most sense for your setup. Security is one area where buying a more expensive hardware product can really help because high-end devices usually include more security features. For example, in some cases you can even find protection for the master boot record (MBR) or GUID partition table (GPT) used to make your hard drive bootable (you can read about the difference between MBR and GPT at https://wiki.manjaro.org/index.php?title=Some_basics_of _MBR_v/s_GPT_and_BIOS_v/s_UEFI and http://www.howtogeek.com/193669/ whats-the-difference-between-gpt-and-mbr-when-partitioning-a-drive/). In order to get GPT functionality, your motherboard must sport a Unified Extensible Firmware Interface (UEFI), rather than the older basic input/output system (BIOS). Some less expensive motherboards lack this feature (you often get what you pay for).

> **NOTE**
>
> *You see many virus reports talking about damage to the MBR. In some cases, the authors have no idea that many modern systems have GPT, not MBR, and that viruses are often specific to one or the other. It's best to assume that any virus out there will attack either GPT or MBR unless you find a report that actually differentiates between the two.*

Creating Useful Passwords

You have probably heard all the password warnings out there at one time or another. The fact is, you need a good password to help thwart the bad guys. This requirement is especially important when it comes to the hardware security features of your system. When someone can gain access to your hardware, it becomes possible to perform all kinds of damaging acts, which can include disabling your computer and turning it into a boat anchor.

When it comes to hardware feature passwords, easy-to-remember and hard-to-guess take on a new meaning. You may not use the password very often, but when you need it, you need it now and not sometime in the future. In some cases, you won't have an easy method of recovering from a lost password either. Writing the password down and putting it into a safe probably isn't a bad idea, but keeping the password attached to the system in some way is as bad as not setting the password at all.

It's important to realize that password strength meters don't necessarily tell you how well your password will resist hacking (see http://www.computerworld.com/article/2902243/many-password-strength-meters-are-downright-weak-say-researchers.html). Use these guidelines when creating your password:

- Avoid using words that are spelled precisely the same as they are in the dictionary. For example, you could replace the letter E with the number 3.
- Make it relatively long and complex.
- Include spaces (when allowed) and special characters (such as the ampersand, &).
- Use a combination of uppercase and lowercase letters.
- Include numbers.
- Misspell a word or two, such as "MiG00dPassphras3".

Passwords that are too hard to remember simply invite people to write them down and place the password where someone else can easily see it. The point is that you want to make things hard on your attacker, but still easy to remember.

Defining the Hazards of WLAN

Chapter 12 provides full security guidelines for WLAN, but it's important to consider the hardware implications of using WLAN for your network. Any time you use radio waves without expectation of eavesdropping, it's akin to shouting information at your neighbor and not expecting the neighbor to hear anything. It's possible to secure a wired connection because the signal remains within the cable. Wireless connections, by their very definition, are open to interception by anyone with the proper equipment.

The reason that alternatives to radio waves exist is because other forms of wireless network connection are at least a little easier to secure and can provide other advantages over radio waves. It's hard to create a directional radio signal that someone can't intercept. It's impossible to secure an omnidirectional radio signal because it fans out in all directions and there is simply no way to avoid having it intercepted. Consequently, when considering your LAN setup, think about the level of security needed. You need to question whether your application can support the risk involved in using a WLAN.

11
Connecting to the Internet

It may seem as if this chapter is coming too soon in the book. After all, it might seem better to install the operating system and applications, and then connect to the Internet because you won't have access to a browser before then anyway. However, most operating systems today rely on an Internet connection during the installation process. The connection lets the installer connect to the vendor site to download operating system updates, device drivers, and other required software to ensure you have a great installation when the installation process completes. In addition, the connection often lets the installer activate the operating system for you so that you don't have to do it manually later.

Of course, the Internet connection won't be functional from a user perspective when you finish this chapter. The purpose of this chapter is to get everything setup and configured so that the operating system installer can access the Internet and perform any required tasks. Obviously, there are limits to the amount of testing you can perform under these conditions and you may find that the Internet connection still isn't available. Most operating systems do provide a fallback position so that you can perform the required tasks later, but doing this configuration now can save you time no matter how things turn out later. With this in mind, the following sections help you set up and configure an Internet connection for operating system use.

Configuring Multiple LANs

If you have a single machine without any Internet connection or already have your home network setup, you can skip this section. However, if you're setting up a home network, then you need to read this section and determine how to configure your Internet connection to work alongside your home network.

Considering the Need for a Router

Whenever you have a local network and an Internet connection at the same time, you actually have two networks. The first network is your local network—the set of machines that is at your home, office, garage, or any other location. The machines may not physically reside in the same location. In fact, chances are good that they don't, but they do represent a single network. The second network is the Internet. The Internet is a completely separate network and you need to treat it as such. In fact, it's a good thing that the Internet is separate because otherwise you'd have trust issues with the machines on your local network.

When you open a browser on your system, you see the Internet. However, when you open a file browser of some sort, you see the local network. Even though the two networks are separate, you access them on the one machine. To some people it might look like there is just one network, but behind the scenes a router sends traffic between the two networks for you. The router acts as a sort of traffic cop—deciding where each packet of data on your system should go.

The router always has two LAN adapters as a minimum. The first LAN adapter connects to the home network, while the second connects to the Internet. Software supplied by the operating system routes traffic between the two connections. You only need one router. If you were to have multiple routers, the networks would become confused because it would be like having multiple traffic cops directing traffic at the same intersection. Instead of preventing collisions, having multiple traffic cops would create them. The same is true of your network—multiple routers would create collisions between the two networks instead of solving them.

In order to provide router services, the machine in question has to be a server. A server is a kind of client application that runs on the machine. The "Considering the LAN Software" section of Chapter 10 tells you about the kinds of LAN software you can install. You have three choices for creating a router for your network:

- Purchase a dedicated router and connect all of your workstations to it. This option has the advantage of providing specialized security features in most cases, a small package, and a moderate cost. The disadvantage is that dedicated routers can be hard to manage.
- Create a special server system and connect the workstations to it. This option has the advantage of offering a flexible environment where you can install other kinds of server applications, such as database managers or an intranet server. The disadvantage is that servers can be expensive.

- Use one of the workstations as a router. This option has the advantage of offering the lowest price, the easiest configuration options, and the fewest number of pieces of equipment. The disadvantages are that workstations can provide slow Internet access and potentially cause security issues.

Connecting to the LANs

When it's time to install the operating system on your new machine, you need to connect your machine to whatever network you plan to use. The operating system requires the connection in order to detect the networks automatically and perform the required setups. You can perform these setups manually, but letting the operating system do it is both easier and faster. In addition, the operating system will likely require the Internet connection in order to download updates and device drivers, and to perform tasks such as activation.

> **WARNING**
>
> *It's important to realize that the standard method of installing the operating system is with the network connections in place. However, there are always risks involved in connecting your machine to the network. One risk is that something, such as a virus, will infect your new machine before the operating system can install the required security updates. The infection can spread to the rest of the local network through your machine. An errant update (made automatically by the installer) could also corrupt the new system. Because you don't have any data on the new system, or even applications for that matter, the outcome of the risk is small, but you do need to be aware of it and perform proper testing after operating system installation just to be sure.*

Determining Your Internet Connection Type

Vendors supply many different methods to connect to the Internet. The connection types available in your area vary by location. For example, if you live a rural setting, your two options might be digital subscriber line (DSL) or satellite. However, when living in a city you might have considerably more options to consider. This chapter can't provide you with a benefits analysis of each connection type because there are too many variables to consider, such as the cost of a particular connection type in a given area. However, Table 11-1 shows the most common Internet connection types and a short description of each.

As you can see, there are a lot of different connection types and this list probably isn't even complete. Vendors come up with new ways of connecting you to the Internet every day. It's up to you to decide issues such as whether you want to pay a

TABLE 11-1 Internet Connection Types

Connection	Reliability	Cost	Speed	Description
Broadband	High	Moderate	Moderate to high	You access this modern Internet connection through a special telephone company, cable, or satellite connection. The speed of the connection depends on the technology used. For example, DSL is only moderately high speed in most cases because many DSL connections rely on the same copper wires used for telephone calls. However, a fiber optic connection would provide extremely high speeds. Broadband is a general term that you need to have defined before you accept service from the provider.
Cable	High: city Low: rural	Moderate	High	This setup uses a cable line to make a digital connection to the Internet. In order to make the connection, you must have a digital MODEM capable of interacting with a cable connection. The connection is extremely fast in many cases, unless the number of subscribers begins to exceed the line capabilities. This form of connection is reliable in a city environment, but may not be reliable in rural areas.
Dial-up	High	Low	Low	This setup uses a telephone line to make an analog connection to the Internet. In order to make the connection, you must have an analog MODEM. In addition, this type of connection ties up the telephone lines so that you can't make a call while using the Internet. You may find dial-up connections in rural areas, but most cities have moved on to other solutions. An advantage of dial-up is that you can use it anywhere where there is a telephone line.
DSL	High	Moderate	Moderate	This setup uses a telephone line to make a digital connection to the Internet. In order to make the connection, you must have a special digital MODEM and your telephone connection box must be set up in a special way to allow the separation of analog and digital signals. A significant advantage of DSL is that you can use it anywhere there is a telephone line, even in cases where no cellular or satellite signal is present. A DSL connection suffers from distance problems, so that it can be hard to get a DSL connection in some rural areas.

Hotspot	Moderate	Low	Low to high	A router with wireless connectivity makes the connection to the Internet and feeds it to any device within the range of its wireless connection. In this case, you utilize Wi-Fi connectivity (see the "Delving into Wi-Fi" section of Chapter 10 for details) to create the connection between devices and router. This setup can be slow (depending on the functionality of the router and the kind of signal used), suffer from dropouts, and create security problems.
Integrated Services Digital Network (ISDN)	High	High	High	This setup normally uses a digital telephone line (as contrasted to the analog telephone lines found in most areas). Instead of a MODEM, you rely on a special ISDN adapter. Both ends of the line must have an ISDN adapter installed.
Mobile	Low	Low	Low to moderate	This is normally an individual Internet connectivity solution. The cellphone or other mobile device provides Internet connectivity through the caller's plan. By creating a connection between the mobile device and your computer, you can use the Internet connection provided by the mobile device. This kind of connectivity doesn't offer anything in the way of security, so you want to employ it with extreme care.
Satellite	Low to moderate	Moderate	Low to moderate	A special MODEM connects to a satellite dish and a signal is beamed between the MODEM and a satellite. It relies on the same kind of technology as satellite television. You normally see this kind of connectivity offered in rural areas. The connectivity can be spotty, especially during storms. Trees and other obstacles can also degrade the signal. In most cases, the download speed is relatively high, but the upload speed is somewhat slow.
Wireless	Moderate	Moderate	Low to moderate	The cables normally used to create connections are replaced with radio waves. In this case, the MODEM includes the hardware required to create a connection from the Internet to any device that needs it. Setting up a wireless MODEM to work with a local network can be tricky.

premium price for a high-speed connection or whether wireless connectivity is worth the drop in connection speed. As the world becomes ever more connected to the Internet, you should plan on finding even more solutions available.

Connecting a MODEM

A modulator/demodulator (MODEM) is a special device to connect a digital device, such as a computer, to an analog line, such as a telephone line. You use a MODEM to connect to Internet sources, such as DSL, that rely on some sort of analog line, generally the telephone system or a private line of the same type. If you're using a connection that doesn't require a MODEM, such as a mobile device, then you can skip this section and go on your merry way toward utter Internet connectivity without bothering with a MODEM. The following sections help you obtain and install a MODEM, assuming that your telephone company or some other third party hasn't done so for you already.

Defining the Purpose of a MODEM

Generally speaking, a MODEM provides a connection between two kinds of signal types (the exact technical details are unimportant for this discussion, but you can read about them at http://computer.howstuffworks.com/MODEM.htm). In the past, you used a MODEM to translate digital signals to analog signals, and vice versa. A common example of such use is the dial-up MODEM used to create connections between computers and various remote servers (not just the Internet, as you might think). However, MODEMs do see other users for other kinds of translation. The point is that a MODEM always translates between one kind of a signal and another.

The reason for the translation varies by the kind of translation performed. However, it's possible to view the need for translation as:

- A requirement to support existing infrastructure
- The need to transmit data long distances without error
- A means to reduce data transmission costs
- A method of piggybacking a data transmission signal onto other kinds of data transmission

Obtaining the Correct MODEM

The term MODEM is a generic—it applies to a wide range of devices. In order to define MODEM better, you must prefix it with another term, such as cable MODEM or DSL MODEM. Even then, the term is nebulous. For example, when buying a dial-up MODEM, you must specify a connection speed, standards it uses, and additional features it supports. Some parameters become standardized over time. For example, you can't buy a dial-up MODEM that supports any speed other than 56 kbps at this point, but at one time you could buy lower-speed dial-up MODEMs at a reduced cost.

> **TIP**
>
> *The best way to avoid problems is to discuss the MODEM connection specifications with your provider to ensure the MODEM you buy is one that will actually work with the provider's setup. Small differences in standards support can cause the MODEM to act erratically or not connect at all, even if you get the right connection speed and meet other requirements. Some providers will rent you a MODEM for a low monthly cost and the reduction in connection headaches is often worth the rental fee you pay.*

In some cases, you must also determine the operating systems that a MODEM supports. An older dial-up MODEM may not support Windows 7, but you need a newer dial-up MODEM to support Windows 8.*x* (and even then, the support may not provide complete functionality on the newer operating system). Support for specific MODEM types varies greatly by operating system, which the Macintosh proving the most difficult to fulfill (unless Apple makes the part you need). Linux support is spotty because each Linux distribution seems to have its own quirks that affect how software interacts with devices' connection to the machine.

It's also important to consider MODEM packaging. An internal MODEM isn't easy to transport between systems, but it also won't lose your connection as easily or require an additional power cord. USB MODEMs need no power, except what the USB port provides, but you often need to find the right USB port and you could find yourself disconnected when someone bumps the machine. External MODEMs packaged in their own case do require additional power, but could have the advantage of supporting more than one system.

Configuring the MODEM

MODEM configuration can be interesting because the amount of configuration you perform depends on MODEM packaging, operating system, and type. A DSL MODEM may not require any configuration at all. You take it out of the package, plug it in, and it's ready for use. A dial-up MODEM may require setting switches before installation, configuration with software after installation, and even some tweaks before each use. Because of the differences in MODEM configuration requirements, this chapter can't provide you with precise configuration steps. The best option is to ensure you view the vendor manual when you get the MODEM to ensure you get a good connection.

Connecting to the MODEM

A MODEM always has two connectors as a minimum. The first connects to the external source, such as a telephone line or a satellite connector. The second connects to the computer. Even when the MODEM is internal and you only see one physical connector, both connections are in place. To avoid potential problems you always follow these steps when connecting your MODEM:

1. Connect the MODEM to the external source.
2. Connect the MODEM to the computer system.
3. Connect the power cable, if necessary.
4. Turn the MODEM on with the computer off to reduce potential damage if you've connected something incorrectly.

What happens next depends on the stage at which you connect the MODEM. If you're just setting your system up, the MODEM won't actually do anything until after you install the operating system and possibly some software (such as a device driver). When adding a MODEM to an existing system, you may see the system perform an automatic setup or request that you provide the vendor software to perform the setup.

Using a Test System to Check the Connection

When you install a MODEM as a prelude to installing an operating system, it pays to check the connections before you start the operating system install. Of course, you can't perform such a check with the target system because it doesn't have an operating system, which means you can't determine whether the MODEM is actually working. The best option is to connect a test system and try the MODEM connection using it.

The basic approach to testing is to ensure you have a usable connection—one that you can use to upload and download data. Make sure you test the connection long enough to ensure that it's reliable. It also helps to test access to some of the sites you think the computer may access during the installation process. Unfortunately, most vendors don't do a good job of telling you what these sites are.

Using Alternative Connectivity Options

You have many other options for creating connection to the Internet. All of these other options are outside the scope of this book because it's unlikely that you'll find them in a home, small business, or other place where a custom-built computer is used. However, you do have high-speed options, such as T1 and T3 to consider. These older connectivity options are expensive and you won't find them in a rural setting, but some people still use them because they provide such high capacity. T1 and T3 lines require advanced technologies, such as a fiber connection. You can read about how T1 connections work at http://computer.howstuffworks.com/question372.htm. The same article discusses other high-speed options that include T3, OC3, OC12, OC48, and OC192.

Considering Internet Security

Any time you connect your computer to any network, you risk some sort of infection. The tradeoff is that you also gain access to resources. Chapter 10 discusses general security for networks and it discusses the need to balance risk with rewards when it comes to security. The following sections describe the special requirements for working through Internet security issues (many of which you should at least consider before you connect your computer to the Internet the first time).

Understanding the Importance of Updates

This book doesn't spend a lot of time discussing updates. Yes, you get some good information about them in Chapter 18. However, updates are extremely important, especially considering how many people and organizations release information about security holes. You may see messages during various phases of the setup and installation process about getting updated software for your system. In all but the most exceptional cases (such as a known flawed update), performing the update is the best way to go. An update will provide you with bug fixes to the software so that hackers will have a harder time gaining access to your new machine.

If you have a second machine and can get to vendor sites to download files, getting updated device drivers and software before you connect your new system to the Internet is a great idea. Burning the new software to a DVD will make the software available whenever you need it to perform tasks on your new system and reduce the potential for attacks before you even get a chance to see your system run the first time.

Defining the Internet Difference

The Internet represents a far bigger risk than the local network when it comes to security. The difference is that no one is actively trying to attack you on your local network. Yes, you can gain an infection from your local network, but it will be due to negligence or other causes, not due to an outright attack. There are people on the Internet who would love to gain access to your system and grab every bit of personal information they can. If you want to keep your data safe, then you need to be proactive about protecting your system and assume that any steps you take are going to be ongoing and not a once during installation issue. The Internet-based attacks change every day, so your strategy for protecting your system must change as well.

Locking Your System Down

Whenever possible, make sure you use good security practices on your system, which means enabling all the hardware-based protection that you can. Most high-end systems today come with great hardware protection features (which is a great reason to avoid that low-cost motherboard and get a top-of-the-line motherboard instead). The security features appear as part of the system setup and they are disabled by default on just about every motherboard available today.

Locking the Browser Down

After you have the operating system installed and install a browser, you should set the browser to restrict every sort of access unless you specifically enable the access for a given site. Setting security this way means that a lot of sites simply won't work when you access them, but it also means you won't accidentally fall prey to a phishing attack or download a virus. Enable only the amount of functionality that a site requires to perform the tasks you need it to perform. Interestingly enough, this strategy also rids you of ads and other unwanted site features in many cases.

> **NOTE**
>
> *Some sites are built so poorly that access by exception, the technique of enabling features only for certain sites, won't work well. In most cases, I personally avoid using these sites and try to find the information I need elsewhere. For me, the risk of infection is too high to give a site carte blanche to my system. However, you need to weigh the site's value to you personally and allow or disallow access based on your risk assessment. Just remember that the next virus to hit your system could trash every last bit of work you've just performed.*

Relying on Firewalls

In addition to hardware security, make sure you have a firewall between your new system and the Internet. In many cases, you can get a firewall as part of a router. Hardware-based protection is the best way to go when you can get it because overcoming hardware-based firewalls is harder than overcoming a software-based firewall. Ensuring your router is fully secured is a good idea. The vendor manual that comes with your router tells you how to access the router, lock the router down, and set useful passwords on it.

Using Other Security Measures

Before you connect to the Internet, make sure you have every security feature you need in place or ready to implement. The reason that security doesn't work for most people is that they don't take time to think it through. However, you wouldn't just leave your doors and windows open while you're gone from home because it's too much of a bother to close and lock them. The same holds true for your computer, locking the door to your computer is an essential part of protecting your investment.

Make sure anyone using your computer is also aware of your stance on security. If necessary, write the rules down and make sure the other party reads them. Children, especially, need to know what the rules are and agree to them before using your system the first time.

Keep your eyes open for major security threats as well. The security threats change all the time and you need to keep on top of them in order to keep your system safe. This practice is akin to keeping track of burglars or other nefarious individuals in your area. You take additional measures when there is a burglar in the neighborhood to keep everyone in your home safe. The same requirement holds true for your computer.

12
Accessing Wireless Devices

Wireless connectivity is an essential part of computer setups today because it provides so much flexibility and greatly reduces costs. The main sticking point with wired connections is that they require someone to install the wires. Even if you install the cabling yourself (and it's certainly doable), there is still the cost of the cabling itself. In addition, some settings don't lend themselves to wired connections. For example, when you rent a building, you don't want to drill holes through the walls in order to run cables from one room to another. Yes, wired connections also have definite benefits (such as higher connection speeds), but many people today view the problems of wired connectivity as daunting.

Unfortunately, wireless connections aren't precisely error free or easy to create. You need to understand how to make wireless connections and ensure that the equipment you use follows the correct standards for your particular setup. In addition, you need to consider the kinds of devices that you connect wirelessly.

Of all the issues you face, security is the biggest problem with wireless connectivity. Broadcasting your data far and wide means that anyone can pick it up. Even when you encrypt the data and take other security measures, you can't be certain that your wireless connection is safe. By definition, using radio waves to transmit data incurs some level of risk and you need to know whether the risks you take are worth the benefits gained. The following sections provide you with the information you need to make the wireless connectivity of your system work safely and securely.

Understanding the Common Wireless Standards

The Institute of Electrical and Electronics Engineers (IEEE) 802 standard has 22 parts that define various levels of network physical level topology—everything from Ethernet (802.3) and broadband (802.7) to wireless (various 802 standards). You may not ever use all 22 parts of the 802 standard. For example, IEEE 802.5 defines the requirements for creating a Token Ring network. A few of these networks probably exist today, but Ethernet dominates the world of computing. With this in mind, Table 12-1 shows the 802 standards that pertain to wireless connectivity.

A Word About Transmission Speeds

When you see that a specification supports a particular data throughput, that's a maximum value. Under the right conditions and with the correct equipment you may see a throughput speed as defined by the specification. However, in reality, you won't normally see the maximum throughput due to any number of issues. Environmental noise, signal blockage (everything from building components to filing cabinets), and available bandwidth (when you share a line with others) are the most common issues affecting throughput, but there are many other issues to consider as well.

If a device determines that it can't broadcast at the maximum frequency, it uses fallback frequencies. For example, the 802.11b standard provides fallback throughput values of 5.5 Mbps, 2 Mbps, and 1 Mbps. In order to determine the true throughput of your connection, you need an application that tests connection speed. One such test application for Internet connections is Speedtest (http://www.speedtest.net/). Because this is a web-based application, you can use it on any device that sports a compatible browser.

The reason that these fallback values are important is that they allow you to continue using the connection, even if connection speed is less than optimal. Unfortunately, when using a fallback value, the throughput is less and your system will slow considerably.

Improving the environment under which the device transmits information will allow the throughput to increase and improve application speed. Considering transmission speed issues when troubleshooting wireless connectivity is extremely important. Throughput is often the culprit when a connection seems to almost, but not quite, work as anticipated.

Ensuring You Have the Correct Wireless Support

The "Understanding the Common Wireless Standards" section of this chapter should tell you something important—it's easy to get a product that doesn't

TABLE 12-1 Wireless Networking Standards

802 Standard	Connection Type	Resource Site	Description
802.11	Wi-Fi (General)	http://searchmobile computing.techtarget.com/definition/80211	Defines the Wireless LAN Media Access Control and Physical Layer specification baseline. The original standard has a number of amendments, such as 802.11a. Any device that is Wi-Fi certified passes the tests required by the pertinent 802.11 standard.
802.11a	Wi-Fi (5 GHz)	http://searchnetworking.techtarget.com/definition/80211a	Defines a version of Wi-Fi that uses the 5-GHz band and provides 54-Mbps throughput. This standard relies on orthogonal frequency-division multiplexing (OFDM), where the signal splits into several narrowband channels and the device transmits each channel at a slightly different frequency.
802.11b	Wi-Fi (2.4 GHz)	http://searchmobile computing.techtarget.com/definition/80211b	Defines a version of Wi-Fi that uses the 2.4-GHz band and provides 11-Mbps throughput. The standard relies on Direct Sequence Spread Spectrum (DSSS), where the signal is broken into small pieces and the device transmits each piece separately. You may also see DSSS as Direct Sequence Code Division Multiple Access (DS-CDMA) and it represents one of two approaches used for spread spectrum modulation.
802.11c	N/A	N/A	There isn't any 802.11c standard now.
802.11d	Wi-Fi (Global Roaming)	http://searchmobile computing.techtarget.com/definition/80211d	This amendment defines the requirements for allowing users of the 802.11a and 802.11b standards to rely on global roaming when using a connection. You see the particulars of this connection enhancement at the Media Access Control (MAC) layer.
802.11e	Wi-Fi (Quality of Service)	http://searchmobile computing.techtarget.com/definition/80211e	Specifies the techniques used to ensure quality of service (QoS) of a connection. This feature is important when performing voice and video transmissions because it helps avoid situations where the transmission appears to stop and start.
802.11f	N/A	N/A	There isn't any 802.11f standard now.

(Continued)

TABLE 12-1 Wireless Networking Standards (*Continued*)

802 Standard	Connection Type	Resource Site	Description
802.11g	Wi-Fi (2.4-GHz Extension)	http://searchmobile computing.techtarget .com/definition/80211g	Defines a higher maximum data rate for 2.4-GHz connections by changing to OFDM. The maximum data rate is 54 Mbps.
802.11h	Wi-Fi (5-GHz Interference Resolution)	http://searchmobile computing.techtarget .com/definition/80211h	The main purpose of this standard is to revolve interference issues suffered by connections that use the 5-GHz connection. In addition, this standard provides enhancements in the form of Dynamic Frequency Selection (DFS) (see http://www.networkcomputing.com/wireless-infrastructure/ dynamic-frequency-selection-why-its-critical-in-80211ac/a/d-id/1234480) and Transmit Power Control (TPC). TPC reduces the power level of the transmitter to the amount necessary to maintain a connection. Using TPC reduces the amount of energy induced into other devices and also increases battery life.
802.11i	Wi-Fi (Security)	http://searchmobile computing.techtarget .com/definition/80211i	Provides additional security for WLAN applications. The security enhancements include more robust encryption, authentication, key exchange, key caching, and pre-authentication.
802.11j	Wi-Fi (802.11a Extension)	http://searchmobile computing.techtarget .com/definition/80211j	Defines Japanese regulatory extensions to the 802.11a specification to allow a frequency range of 4.9 GHz to 5.0 GHz.
802.11k	Wi-Fi (Resource Measurement)	http://searchmobile computing.techtarget .com/definition/80211k	Specifies techniques for making radio resource measurements for networks using the 802.11 family of specifications.
802.11l	N/A	N/A	There isn't any 802.11l standard now.
802.11m	Wi-Fi (Documentation)	http://searchmobile computing .techtarget.com/ definition/80211m	Provides maintenance of the 802.11 family of specifications by correcting and amending existing documentation.

802.11n	Wi-Fi (Speed Enhancement)	http://searchmobilecomputing.techtarget.com/definition/80211n	Defines higher-speed transmission standards with speeds of 108 Mbps, 240 Mbps, and 350+ Mbps. These competing technologies from EWC, TGn Sync, and WWiSE tend not to be interoperable. The purpose of this standard is to create a common method for performing the speed increase and to increase reliability as well. This standard relies on multiple input, multiple output (MIMO), where multiple antennas at the sender and receiver combine to form a single data transmission path.
802.11o through 802.11w	N/A	N/A	There isn't any 802.11o through 802.11w standard now.
802.11x	N/A	N/A	A misused generic term for the 802.11 family of specifications.
802.15	Wireless Personal Area Networks	http://searchmobilecomputing.techtarget.com/definition/80215	Defines a method for creating a Wireless Personal Area Network (WPAN).
802.15.1	Bluetooth	http://searchmobilecomputing.techtarget.com/definition/Bluetooth	Defines a short-range wireless technology for devices such as cordless mice, keyboards, and cordless headsets. The standard relies on a 2.4-GHz connection. This specific can also support data transfer needs at distances of 10 m or less.
802.15.2	N/A	N/A	There isn't any 802.15.2 standard now.
802.15.3a	UWB	http://whatis.techtarget.com/definition/ultra-wideband	A short-range ultra wideband (UWB) link similar to Bluetooth that allows transmission up to 230 feet with little interference. This technology can work through doors.
802.15.4	ZigBee	http://searchmobilecomputing.techtarget.com/definition/ZigBee	A short-range, low power, mesh network designed for use in a wide area (see http://searchnetworking.techtarget.com/definition/mesh-network for a description of mesh networks). Devices that rely on this standard are often industrial applications or battery-powered Internet connection devices. This specification uses a 2.4 GHz, 900 MHz, or 868 MHz connection.

(*Continued*)

TABLE 12-1 Wireless Networking Standards (*Continued*)

802 Standard	Connection Type	Resource Site	Description
802.15.5	Mesh Network (Extension)	http://searchnetworking .techtarget.com/ definition/mesh-network	Provides an extension of network coverage without increasing the transmission power or receiver sensitivity. The goal is to provide enhanced reliability by increasing route redundancy (giving the signal more paths to travel). In addition, this specification provides methods for making networks easier to configure and to improve device battery life.
802.16	WMAN	http:// searchmobilecomputing .techtarget.com/ definition/80216	Provides a family of standards that defines both fix and mobile broadband wireless access methods that create a Wireless Metropolitan Area Network (WMAN). Any device that implements this standard fully and passes certification testing receives a WiMAX certification. The standard relies on ODFM. Specifications in this standard rely on an unlicensed (900-MHz, 2.4-GHz, 5.8-GHz) or licensed (700-MHz, 2.5- to 3.6-GHz) connection.
802.20	Mobile Broadband Wireless Access	N/A	This is a placeholder for a future standard.
802.22	Wireless Regional Area Network	N/A	This is a placeholder for a future standard.

Chapter 12: Accessing Wireless Devices

provide the standards support you need. Unfortunately, standards support, while complex, isn't the only consideration you need to make. The following list tells you about the sorts of things you need to consider:

- Specific standards adherence: When obtaining products for your wireless network, every item has to adhere to the specific set of standards you want to use. Trying to mix an 802.11a and an 802.11b device will never work.
- System connectivity requirements: It's essential to get devices that actually match the capabilities of your system when it comes to connectivity. For example, some network devices require a USB 3.0 port. If all you have are USB 2.0 ports, the device will never work.
- Device software compatibility: The software available for the device has to match your system and operating system. It's important to read all the fine print about how the device, software, system, and operating system all interact from a software perspective.
- Environmental compatibility: Trying to create a wireless network in a room with lots of obstacles or in a multi-floor scenario requires that you take extra time in selecting devices. In this case, even if an 802.11a or 802.11b device would normally work, you probably want devices that adhere to the 802.11n standard to ensure maximum throughput.
- Power considerations: In some cases, you need to consider the output power of the devices you use. Some people have this "more is always better" mentality that simply doesn't work when dealing with some types of wireless situations. Using devices that output too much power can cause all sorts of interference problems. Think about a restaurant where everyone is talking at the same time—quite loudly. It can become quite hard to hear the person talking next to you because of all the interference; much less hear someone from across the room. However, if everyone talks quietly, then hearing other parties becomes easier. Likewise, when devices output too much power it becomes hard for any one particular device to hear another. More power is helpful when creating long-distance connections or dealing with specific wireless conditions. Smart power allocation is the main reason for standards such as 802.11h.

Configuring Common Computer Devices

Many wireless LAN adapters are self-configuring. All you do is plug them in and they're ready to go. When your computer recognizes the new device, you may need to provide a special disk containing the device driver, but that's normally the end of the configuration process. In most cases, LAN adapters are one of the few situations where a generic driver can work as well as a vendor-specific driver. Chapter 14 discusses device drivers in more detail, but in most cases, the vendor-specific driver won't improve throughput. What you will often get is access to status information and the vendor may also provide a diagnostic you can run if the LAN adapter should fail to work as anticipated.

In some situations, you may have to provide some configuration information for your LAN adapter, but the configuration is network-oriented, rather than device-oriented. For example, when you have multiple LAN adapters in a single machine, you often need to configure the particulars of each LAN and decide how they should interact (if they need to interact at all). This kind of configuration is operating system–specific, so you need to locate the instructions for your particular operating system (and operating system version).

The one configuration task you must perform for all wireless devices, no matter what their use might be, is wireless security. This issue appears in the "Considering Wireless Device Security" section of this chapter. Unless you want everyone in the world to access your wireless access point (WAP) for their own personal gain, you really do need to take extra precautions when dealing with wireless security. The better you configure the security features of any wireless devices you attach to your network, the fewer security issues you'll experience.

Configuring Alternative Devices

Given the purposes for which you might build a custom system, you may want to connect to devices other than other computers, routers, and the Internet. For example, you may want to connect to cameras, alarms, and other sorts of devices. The same rules apply to these other devices as they do to other computers. However, in this case, you may have to obtain LAN adapters that suit the device's needs, rather than the other way around. For good communication to occur, the device and its associated computer must speak the same language in the form of standards and wireless capabilities.

Alternative devices will most likely require configuration. For example, when working with certain kinds of sensors, you must configure the device, apply power, and then bring it within close proximity of the LAN adapter to create a connection. After that, the device could be quite some distance away and still maintain connectivity. (I've personally tested sensors up to 1000 yards away.)

Many alternative devices feature sensitivity settings that you have to configure. In addition, each device will likely require its own channel. The vendor documentation that comes with the alternative device will tell you about the specifics, but be prepared to experiment a little to obtain the best results.

Chapter 12: Accessing Wireless Devices

> *TIP*
>
> *Never assume that an alternative device problem is exclusively due to the network connection or the installed software. A sensor setting may affect how the device works and some devices come with relatively complex settings. For example, a proximity sensor may require that you configure the scan time differently for particular kinds of driveway usage in order to detect passing cars. Otherwise, a fast moving car could possibly move right past the sensor between scans, leading you to believe the sensor is dead.*

Depending on where you install an alternative device, you may also have to compensate for the effects of weather, vibration, and other potential sources of error. It's quite possible that a device is fully functional, but a weak battery keeps it from communicating. Vibration does really strange things to sensors (such as resetting the device when the battery bounces in the holder), so you need to consider all potential sources of problems when creating the wireless connection.

Considering Wireless Device Security

Security will likely consume more of your time than any other aspect of working with wireless devices. The problem is that there are so many different ways to attack a wireless signal, so many standards to deal with, and that the industry has had so many false starts in dealing with security. No one really considered just how hard it would be to secure a wireless signal at the outset. Just getting the technology to work seemed hard enough. The following sections provide you with a good overview of wireless device security.

> *WARNING*
>
> *There is no way to secure a wireless connection completely. You must always assume that someone will gain access to your network, should they want to do so. Even if you don't have any data that you think someone might want to steal, they may not actually be after your data. They may not want to corrupt or damage your system either. They may be spending time with your wireless setup to send out streams of spam e-mails or perform other nefarious tasks using your connection. The authorities will blame you for anything that happens with the connection, so it's actually in the perpetrator's best interest to maintain a low profile. When using a wireless setup, you must take a proactive stance to security, assume someone is using the connection illegally, and be prepared to track that usage down. Yes, it sounds a bit paranoid and, unfortunately, being a little paranoid is a good thing when it comes to wireless security.*

Defining the Types of Intrusions

All sorts of threats plague wireless security. Some of these threats are the same as those you deal with when working with wired security, but many others are specific to wireless security. The following list provides a brief description of common threat types. In order to protect your wireless setup, you must be prepared to handle all of them.

- **Accidental association:** You and a neighbor both have a wireless network. Every time you turn on your computing device, it seeks out the strongest signal it can find, which could be your neighbor's network, instead of your own. Without appropriate security in place, you might never know that you've connected to your neighbor's network, but you now have access to anything the neighbor does, such as financial information stored on a network drive.
- **Ad hoc network:** Another computer connects directly to your computer, rather than to a WAP. The peer-to-peer connection is allowed, by default, when working with some operating systems, such as Microsoft Windows. You must specifically disable it. If you don't disable it, the person connecting to your system may have access to all your personal information, along with any information you can access through a network connection.
- **Denial of service (DOS):** A denial of service attack occurs when a perpetrator keeps making requests of a server and eventually overloads it. If the attacker can get the server to fail in a specific way, it's possible to gain access to the network, perform some reconfiguration, and gain permanent access when the server is brought back online. Even if the perpetrator doesn't gain access, using a DOS attack can inconvenience an organization by bringing the network down or expose potential security issues that a perpetrator can use to gain access.
- **Identity theft (MAC spoofing):** Someone sits in the parking lot of your organization or simply out on the street in your neighborhood and monitors network traffic. The idea is to detect the MAC address of a computer that has network access. When the computer that really has the correct MAC address goes offline for some reason, the individual uses software to spoof the MAC address and gain access to the network. This technique only works well in situations where the computer with the real MAC address goes offline at predictable times. Two computers can't have the same MAC address on the network, so the perpetrator needs to know that the MAC address is free.
- **Man in the middle:** Viewing your hard drive data can be fun for some attackers, but many want to gain access to your personal information. In order to do this, they often need more access than just your network

provides. Once an attacker gains access to your network, it's possible to configure a strong temporary access point. Your computer logs into the strong access point provided by the attacker and the attacker connects to your WLAN through your regular WAP. You can't tell any difference because the network travels just as it did before with the attacker's computer in the middle. However, now the attacker can see everything you're doing, such as entering passwords that are later used to gain access to your personal information. This is just one form of man in the middle. There are many other ways to perpetrate this kind of attack, most of which are completely unnoticeable to the attack target.

- Nontraditional networks: All sorts of devices connect to a network using all sorts of technologies. For example, a handheld scanner can use a Bluetooth connection to eventually gain access to a network. You can't secure Bluetooth connections, so it becomes easy for a third party to gain access to the network through the handheld scanner. There are all sorts of devices in this category: printers, copiers, PDAs, and barcode readers are just a few of the potential candidates for enabling intrusion to a network. You must secure these devices by keeping them from direct network access and verifying that any device they do interact with remains intrusion free.
- Operating system–specific: Some operating systems have specific flaws that make it possible to gain network access. For example, the café latte attack (so named because it only takes 6 minutes to perform—the same time it takes to sip a café latte) relies on problems with the Windows wireless network software. By sending Address Resolution Protocol (ARP) packets to the network, the perpetrator can eventually guess one of the keys used to access the network using statistical analysis. The point is that you need to be aware of the fact that your operating system could provide access to your wireless LAN, so keeping things patched is essential. You can read more about the café latte attack at http://www.wi-fiplanet.com/tutorials/article.php/3716241.

Protecting the Communication Channel

In dealing with wireless security, you must first secure the communication channel. Wireless devices could use any of a number of methods to perform this task. However, the three levels of protection currently recognized by the 802.11 standard are:

- Wireless Equivalent Privacy (WEP): The original form of encryption provided for wireless devices. As originally implemented, WEP was incredibly easy to break because it used such a short encryption key. Longer encryption keys helped, but the encryption technique itself is flawed and as computing devices gain in power, it becomes easier and easier to break WEP. You can read more about the issues with WEP and discover why you should use WPA instead at http://www.howtogeek.com/167783/htg-explains-the-difference-between-wep-wpa-and-wpa2-wireless-encryption-and-why-it-matters/.

- **Wi-Fi Protected Access (WPA) Level 1:** As criticism of WEP mounted, various vendors created a new standard, WPA, that has a significantly longer encryption key and overcomes some of the issues with the WEP encryption algorithm. In addition, WPA contains checks that help detect whether a breach has occurred, making it possible to recover from outside interference in at least some cases.
- **WPA Level 2 (WPA2):** Hackers eventually overcome any wall put up against them and that's why there is WPA2—it's essentially a higher wall. This encryption technique uses a better algorithm and improved communication protocols. The encryption key has also increased in size (yet again). Unlike WPA, some hardware can't upgrade to WPA2 without a firmware update. Your best bet is to ensure every wireless device you get has WPA2 security in place by default.

> **WARNING**
>
> *Avoid the use of Wi-Fi Protected Setup (WPS). This feature was added to Wi-Fi in order to reduce the amount of work required to configure a home network. The problem is that the Personal Identification Number (PIN) feature of WPS is enabled on many devices by default. Because the PIN has a limited number of unique combinations, someone can use brute force techniques to guess the PIN relatively quickly, which then allows access to the device and ultimately the network as a whole. You can read the details about this particular issue at http://www.howtogeek.com/176124/wi-fi-protected-setup-wps-is-insecure-heres-why-you-should-disable-it/.*

Protecting the Network

The communication technique is only one avenue of attack for wireless networks. It's possible for the network itself to become a source of potential intrusion. For example, someone could add another wireless access point (WAP) in order to gain unauthorized access to the network without actually doing anything to existing routers or devices. The following list describes two wireless network protection features:

- Wireless Intrusion Prevention System (WIPS): The basic idea of this technology is to detect unauthorized WAPs and then use countermeasures to disable them. The idea is to keep a wired LAN free of any wireless connectivity and therefore free of potential sources of wireless connectivity intrusion. This sort of technology is often used in larger organizations where employees can bring a wireless router from home and connect it to the network. However, the technique is also used in any situation where you feel someone could compromise an existing network by adding unauthorized WAPs.
- Wireless Intrusion Detection System (WIDS): A WIDS goes a step further than a WIPS and detects intruders on both wired and wireless networks by checking for spoofed MAC addresses and other signs that an outsider has connected to the network. Every device on a network has a unique MAC. Devices with a spoofed MAC tend to exhibit electrical characteristics and other clues that the device isn't actually part of the network.

Understanding the Role of Wireless Access Points

Some organizations seek to keep their wireless connections more secure by eliminating WAPs. Devices connect directly to each other as needed. Unfortunately, the devices are still communicating with each other using wireless technology and the signal is still open to outside interference and interception. Anyone could sit in close proximity (such as a parking lot or on the street) and gain access to the wireless signal. Not using WAPs doesn't do anything for security on your wireless network, despite rumors you may have heard to the contrary.

WAPs do help you enforce security. You can configure a WAP to require specific logins and to enforce security policies. By forcing everyone to use a WAP, you make connectivity to a wireless network consistent and reliable. Consistent access is especially important because it makes access predictable. When working with wireless networks, anything you can do to reduce the number of uncertainty factors is good.

Part IV

Installing the Software

13

Installing the Operating System

You have a fine system put together at this point. All the hardware that you could ever want is in place. Unfortunately, hardware can't really do much by itself. In order to obtain a functional system, you need an operating system. The operating system provides the environment in which applications work. Most people ignore the operating system to an extent, but the operating system really is there in more ways than you might imagine.

This chapter discusses operating systems. It doesn't specifically tell you how to install any particular operating system because there are instructions for doing that online. Instead, the chapter helps you understand the process for installing an operating system and helps you focus on issues that you won't find online—such as how to choose an operating system that really will fulfill your needs. You also find some ideas on how to get around installation issues that plague the person building their own system, such as how to get the Mac OS X to cooperate (assuming you want to install it).

Choosing an Operating System

Before you can install an operating system, you need to choose one. You may have had an operating system in mind before you began building your new system. However, things may have changed in the meantime. Perhaps your needs have changed or technology has moved on. The following sections help you think through what you want for an operating system. You may decide that you really do want something different from what you originally thought you'd like.

Any Operating System Will Do

Operating systems aren't magical. They have no special powers and you don't have to perform weird incantations to get them to work. Like all software, operating systems are instructions that tell the hardware what to do. It's just like any list of instructions you create for any other purpose. The list defines the outcome of whatever it's designed to control. The operating system you install will define the manner in which your system interacts with applications and defines the functionality you can expect your system to provide.

The main criterion for an operating system is that it speaks the same language as the hardware. This may seem like an odd requirement, but different processors speak different languages. If your operating system doesn't speak the right language, the hardware won't listen. This is the reason you could potentially still use the Disk Operating System (DOS) on an Intel processor system that you build. DOS still speaks the language of the processor, even though it won't use even a smidgen of the capabilities that a modern machine provides. Windows 95 will still install on a system, but that's a security breach just waiting to happen. The point is that any operating system that speaks the Intel language will do as your operating system.

Defining What You Want to Do

Different operating systems are better at particular tasks. For example, many people like to use Linux for experimenter setups because Linux provides better control over the environment and there is a large assortment of scientific software available for it. Many people rely on Windows because it has a huge installed base, runs business software and games relatively well, and provides great device driver support. Mac OS X often appeals to people with artistic needs. Of course, these are all generalizations and you can find examples of software that runs on every one of these operating systems.

Part of your decision will hinge on what you can afford and how much effort you're willing to employ to obtain a particular result. You can get a copy of Linux free. Windows will cost you something, but it's generally affordable for most people. Even though you can get a copy of Mac OS X, it won't install on your custom-built PC without a lot of extra effort. Then again, you might be willing to put in the required effort to achieve your goals.

Operating systems also vary by ease of use. Many people consider Mac OS X the easiest of the operating systems to use because it provides an intuitive interface. Windows comes next in line. The flexibility that Windows offers comes at the price of added complexity in the interface. Linux can be quite difficult to use, even though modern versions of Linux come with a GUI.

Another issue is device support. Windows is the strongest operating system in this area. Because it has a huge installed base, most vendors write device drivers

Chapter 13: Installing the Operating System

for it first. Mac OS X usually comes next with good device driver support. Linux can be a bit patchy with device driver support because there are so many different flavors of Linux to support. You want to be sure that whatever flavor of Linux you choose supports the devices you chose to install in your system. The important point of this section is that you need to explore each of the operating systems enough to know that the operating system you get will actually fulfill your needs and work with the hardware you have installed.

Considering Version Issues

The version of an operating system you obtain can make a big difference. For example, Microsoft has a spotty history of producing really good operating systems. Read about either Windows Vista or Windows 8 and you find that many people feel they are the worst operating systems that Microsoft ever put together. (Just which one is worst is a matter of much debate.) Of course, you might disagree with the crowd and find that you love Windows 8. The point is that each version of Windows has significant differences with others in the series, so you have to choose carefully.

You can consider the flavors of Linux as a type of version decision. However, with Linux, the differences are even more significant. When considering a Linux setup for your system, you need to think about:

- Distribution: The specific kind of Linux you want to own, such as Ubuntu or Debian.
- Desktop environment: The user interface you employ while working, such as Unity or GNOME.
- Version: A specific version of the Linux distribution and desktop environment that you choose.

At this point, your eyes are probably glazing over if you're thinking about installing Linux. Fortunately, there are sites where you can find comparisons of the various distributions and desktop environments. You can find a listing of the most popular distributions (along with a review) at http://www.howtogeek.com/191207/10-of-the-most-popular-linux-distributions-compared/. A listing of the most popular desktop environments appears at http://www.howtogeek.com/163154/linux-users-have-a-choice-8-linux-desktop-environments/. After you choose a distribution and desktop environment, choosing the versions you want to use is a lot easier.

When choosing Mac OS X you normally don't have many options. Unless you're willing to scope out eBay looking for an old version of Mac OS X, you usually have the option to buy the latest version and that's about it. In some cases, you can find the previous version as well, but Apple is pretty consistent about limiting choice. Currently you can buy copies of OS X 10.6 (Snow Leopard) or OS X 10.7 (Mountain Lion). In order to get one of the newer versions of OS X

(Mavericks or Yosemite), you must first install Mountain Lion and then download the newer version from the Apple store. You can find a comparison of the various OS X versions at http://apple.wikia.com/wiki/List_of_Mac_OS_versions

Considering Longevity

Software doesn't wear out. It does get old, but you can continue using it for as long as you like and the instructions will continue to work just as they always have. The longevity of a piece of software isn't measured by the vendor's upgrade cycle. Instead, you measure the longevity of software by its ability to meet your needs. As long as the software continues to meet your needs, then you can continue using it, just as you always have. This fact seems lost on many people who seem to think that they must upgrade the second a vendor comes out with a new version of their product. The marketing people love it when you think that way, but when you upgrade to a new operating system depends solely on your needs. Instead of listening to marketing, consider these requirements as part of the longevity calculation for your operating system purchase:

- Security: If you plan to connect your system to the Internet, then the operating system is good only as long as you can keep it secure. Operating system vendor bug fixes are important, but they're only one aspect of keeping your operating system secure. Using third-party products, such as antivirus and firewalls, can help a great deal. Even so, at some point security threats will become impossible to ignore and you either have to disconnect from the source of infection (the Internet) or upgrade your operating system.
- Hardware support: Even though new hardware is becoming less of a problem, the operating system is only good as long as you have hardware that supports it. When you start encountering devices that don't provide device drivers for your operating system, you need to choose between continuing to use older hardware that does work or upgrading the operating system.
- Applications: As with an operating system, you don't have to upgrade your applications constantly. As long as your applications do the job they're supposed to do, you can continue using them. Even with items that you use to interact with other people, such as a word processor, there are usually solutions for continuing to use a version of the application that your operating system supports. However, at some point, you may find a version of an application you really do need and it won't work with your current operating system, so you'll need to upgrade.

Just how long an operating system can last depends on you. I currently have a system I built out of all the old parts that no longer work for my main system.

The system runs Windows 98 and it has an old version of Office running on it. I use the system to play older games and to maintain various personal databases. This setup also offers me access to DOS, which I use for various purposes, such as running old games and other ancient software. It's an interesting system because some of the hardware was created well after Windows 98 was no longer a contender for any sort of business need. Of course, I have the good sense not to connect this system to the Internet. It's a stand-alone system in my family room and used only for personal needs.

Exploring Alternative Solutions

Your PC hardware can run a huge number of operating systems. Just because this book focuses on the main three operating systems, Windows, Linux, and OS X, doesn't mean that you have to install them on your new hardware—quite the contrary! You have access to an amazing range of alternatives. Here are just a few of the more interesting choices (in order of level of appeal and functionality):

- SteamOS (http://store.steampowered.com/livingroom/SteamOS/): Gamers know that many of the coolest games available today run under Steam. However, if you're using a PC, you have to jump through some hoops to get everything up and running. Installing SteamOS means that your system is always ready to play games. This is another kind of limited Linux distribution, but it's designed specifically for gamers.
- Haiku (https://www.haiku-os.org/): You might remember BeOS from day gone by (read about BeOS at http://www.theregister.co.uk/2007/01/30/forgotten_tech_beos/). Haiku is a BeOS implementation that shows where BeOS might have been today had Microsoft not stepped in with its usual heavy-handed swat of the competition. This operating system has some truly interesting features and it might be a good choice for the experimenter or someone who performs specialty tasks, such as security systems. Because many of the attacks that work fine on Windows won't work on Haiku, you avoid many of the security risks that other people have to face.
- ReactOS (https://www.reactos.org/): Many people like older versions of Windows based on the Windows NT operating system. ReactOS is an open source emulation of Windows. It doesn't use Windows code, but it does work with Windows drivers and applications. It's an operating system to consider if you like the older Windows NT-style interface and dislike the direction Microsoft is taking with Metro.
- Chrome OS (http://www.chromium.org/chromium-os): If you only intend to use your new PC for browsing resources on the Internet, you might want to load the same operating system that a Chromebook uses. You can use this operating system to run various Chrome apps and it

might make for a good entertainment-type system. Chrome OS is a kind of limited Linux distribution.
- eComStation (http://www.ecomstation.com/): OS/2 was one of those interesting operating systems that Microsoft ended up killing. It actually fixes many of the problems that exist in Windows and has a somewhat large installed base of applications available for it (see http://www.ecomstation.com/product_info.phtml?url=nls/en/content/tour-software.html for details). IBM continued to support OS/2 for a long time after Microsoft pulled out of the development effort, but eventually failed to keep OS/2 going. eComStation is the latest implementation of OS/2. This would be an excellent operating system for business users, scientists, and experimenters. The options for gamers are somewhat limited, but it's the platform you want if you want to be able to play older games.
- Syllable (http://web.syllable.org/pages/index.html): If you liked the Amiga and were sorry to see it go, this would be a good operating system to look at. The interface looks like Amiga in some ways, but also adds some Linux elements. This would be an interesting operating system for experimenters.
- FreeDOS (http://www.freedos.org/): Some people still need a DOS-like environment to run really old games and work with other older software. Experimenters sometimes need an operating system with a really small footprint to perform embedded systems work. FreeDOS is precisely what the name says, a version of DOS that you can download for free. It supports the usual character-mode interface and you can use it as a starting point for installing older Windows versions, such as Windows 95.
- Android (http://www.android.com/): This is one of the options where you wonder why you'd want to install it in the first place. This is a Linux distribution of a sort, but it's so limited you have to wonder why someone would install it unless they're addicted to their Android. Even so, if you install this operating system, you do gain access to all the Android apps and can run them one at a time (that's right, this operating system doesn't support multitasking).

Obtaining the Vendor Installation Instructions

Before you can use an operating system, you need to install it. Trying to fit just the Windows installation instructions in a single chapter would be impossible (they might fit well in a book), so the following sections don't include step-by-step instructions. The following sections do provide you with some resources you can use to install a copy of your favorite operating system.

Chapter 13: Installing the Operating System

Locating the Windows Instructions

Microsoft changes the Windows installation adventure with each new version of Windows. The installation instructions that worked great with the last version won't work with the latest version. With this in mind, the following list provides links to installation instructions for specific versions of Windows:

- Windows 95: https://support.microsoft.com/en-us/kb/149712
- Windows 98: https://support.microsoft.com/en-us/kb/221829
- Windows 2000: https://support.microsoft.com/en-us/kb/304868
- Windows ME: http://pc.buildyourown.org.uk/installing-windows-me/
- Windows XP: https://support.microsoft.com/en-us/kb/978307
- Windows Vista: http://windows.microsoft.com/en-us/windows/installing-reinstalling-windows#1TC=windows-vista
- Windows 7: http://windows.microsoft.com/en-us/windows/installing-reinstalling-windows#1TC=windows-7
- Windows 8/8.1: http://windows.microsoft.com/en-us/windows-8/clean-install

You may have noticed that Windows NT 4.0 is missing from the list. The instructions for this version of Windows are so convoluted and difficult that vendors often put out their own custom guides for installing the operating system. Microsoft doesn't actually provide a guide for installing Windows NT 4.0 anymore, but you can find a number of support articles that could help out such as:

- INFO: Windows NT 4.0 Setup Troubleshooting Guide (https://support.microsoft.com/en-us/kb/126690)
- How to Create Windows NT Boot Floppy Disks (https://support.microsoft.com/en-us/kb/131735)
- Windows NT 4.0 Driver Library (https://support.microsoft.com/en-us/kb/142643)

Choosing a Linux Flavor

Linux installations tend to differ by distribution. Yes, you can probably install any version of Linux once you know how to install one of them, but it's best to follow the instructions for that particular distribution to ensure you get the best results. Table 13-1 lists the most popular Linux distributions with the location of the installation instructions for that distribution. Because you may also need additional information, such as how to change the desktop environment, the table also includes a link to the main page.

TABLE 13-1 Linux Installation Locations

Linux Distribution	Main Page	Installation Instructions
Ubuntu	http://www.ubuntu.com/	http://www.ubuntu.com/download/desktop/install-ubuntu-desktop
Linux Mint	http://linuxmint.com/	http://www.wikihow.com/Install-Linux-Mint
Debian	https://www.debian.org/	https://www.debian.org/releases/stable/installmanual
Fedora	https://getfedora.org/	http://docs.fedoraproject.org/en-US/index.html (choose a version number in the window on the left and the Installation Guide link below that)
CentOS	http://www.centos.org/	https://www.centos.org/docs/5/html/Installation_Guide-en-US/s1-installmethod.html
Red Hat Enterprise Linux (RHEL)	http://www.redhat.com/en/technologies/linux-platforms/enterprise-linux	https://access.redhat.com/documentation/en-US/Red_Hat_Enterprise_Linux/ (choose a version number in the window on the left and the Installation Guide link in the Getting Started topic)
openSUSE	https://www.opensuse.org/	https://en.opensuse.org/Portal:Installation
SUSE Linux Enterprise	https://www.suse.com/	https://www.suse.com/documentation/ (choose a version link from the list, then the Deployment Guide link from the Installation and Administration category)
Mageia	https://www.mageia.org/	https://wiki.mageia.org/en/Installing_Mageia
Arch Linux	https://www.archlinux.org/	https://wiki.archlinux.org/index.php/Installation_guide
Slackware Linux	http://www.slackware.com/	http://www.slackware.com/install/
Puppy Linux	http://puppylinux.org/	http://puppylinux.org/main/Overview%20and%20Getting%20Started.htm

Hacking OS X

Apple doesn't want you to install OS X on anything but real Mac. However, your hardware will run OS X just fine if you can get past the obstacles that Apple has put in your way. What you end up with is a Hackintosh (as contrasted to the Macintosh you get when you buy the hardware from Apple). Of course, someone is always willing to figure out ways to get around problems, especially when you tell them they can't do it. That said, creating a Hackintosh does involve all sorts of things that this book can't truly recommend you do—you need to decide whether you're willing to live with the consequences. All this said, you can find complete instructions for creating the Hackintosh of your dreams at http://www.hackintosh.com/. You can find step-by-step instructions and tutorials for building a Hackintosh sporting the following versions of OS X:

- Yosemite
- Mavericks
- Mountain Lion
- Lion
- Snow Leopard
- Leopard

Ensuring You Have All the Details

By the time you get to this point in the chapter, you know that getting an operating system installed on your new system could be a daunting task. People usually fail to get a good installation because they fail to plan carefully. What you really want is an organized approach in mind before you begin the installation. The following sections describe the details you should have in place before you begin installing the operating system on your new machine.

Getting a Licensed Copy

Windows and OS X are two examples of operating systems you must buy before you install them. A few of the Linux installation also require that you buy before you install. Yes, you can get free versions of just about any software from an online download site. However, these supposedly free versions almost always come with a price. When you install this software, you also get free software such as viruses, Trojans, back doors, and adware (and that's only if you're lucky). In general, when someone offers you a free copy of software you'd normally need to buy, there is a reason they're giving it to you and it isn't in your best interest to accept.

When you do choose a free operating system, such as Linux Mint, be sure to download your copy directly from the support site. When you use an alternative site, you again take the risk that the copy you get will provide you with extra bonuses that you really don't want to install on your new machine.

No matter whether your operating system is paid or free, you need to observe the licensing details. Many people get products without ever reading the fine print, but it really does pay when working with your PC. For example, some software you install gives the provider permission to do things like poke around on your hard drive and upload information about you to the vendor site without your permission. You have to ask yourself whether it's ever a good idea to give someone too much control over the machine you've just put together, no matter how appealing the software might be.

Obtaining Manuals

Operating systems come with manuals. In fact, most software comes with a manual of some sort. It's true that the manual might not be worth much, but it usually provides useful information that might save you considerable time and effort. Downloading and at least scanning the manuals before you begin an installation are essential. Make sure you check any requirements and needed equipment carefully. The documentation might also include a list of known issues that you need to know about before you proceed.

> **TIP**
>
> *You might think that having the manuals online is good enough. It's quite possible that your installation process could bring your network down or make it impossible to access the Internet for some other reason. Always assume that you won't have required connectivity when you perform an installation of any sort. Having the manual on a local drive that you know you can access during the installation or even printing out essential sections is a good idea.*

Performing Backups

Creating a backup of your data, all of your data, before you start an installation is important because you never know when an installation will go horribly wrong and take your data with it. Before you perform any sort of installation, especially an operating system installation or upgrade, make sure that you have a copy of your data in a location other than a hard drive. Tape, DVD, or external hard drive (removed after the backup is complete) backup is the best way to go because you don't want anything to be able to access the backup during the installation process.

Chapter 13: Installing the Operating System

Downloading Device Drivers

Most operating systems today provide generic device drivers that you can use in place of vendor-specific device drivers (see Chapter 14 for a discussion of device drivers in detail). The vendor-specific device drivers generally provide better functionality and higher reliability than the generic device drivers do. It always pays to have the latest copies of any device drivers needed for the devices you installed into your system to ensure that you can complete an operating system installation without problem.

Although it has become rarer, you may actually need a device driver to access the hard drive on your new system. The operating system will usually request the device driver when it determines that hard drive access is impossible. However, sometimes all you see is a prompt telling you that you can provide a device driver if needed. Of all the device drivers for your system, any required hard drive device drivers are the most important. You can replace most other generic device drivers later using the techniques described in Chapter 14.

> ### WARNING
> *Never assume that the device drivers supplied on a disk with the device are the latest drivers. Always download the latest drivers before you begin the installation to ensure that the installation is as error free as possible. Newer device drivers often enhance device functionality, fix security issues, and reduce the potential for bugs causing reliability problems.*

Performing the Installation

At some point, you've made all the required choices. You know which operating system you want to install, have all of the required resources in hand, and are certain that you have installed the hardware correctly. Installing the operating system shouldn't be hard. The biggest thing to remember is to be patient and take your time.

After you perform the initial installation, you may need to perform some required configuration tasks. The operating system has to perform these tasks to ensure the environment is stable and usable when the installation process completes. However, most installers leave a bit of additional work for you to do. Here are some things to consider checking before you start installing any applications.

- Verify that all the operating system features you want are actually installed.

- Ensure that all of your devices are configured and operating (see Chapter 14).
- Check all your network connections to ensure they work.
- Access the Internet (if required) to ensure you can perform installation completion tasks.
- Activate the operating system if your operating system has an activation feature.
- Reboot the system to ensure any changes you make take hold.

14
Accessing the Devices

Attaching devices to your system isn't some magical process where connections just occur and things simply work. Yes, vendors have told you that very thing for years, but it isn't true. A lot of work takes place under the seeming magic to make the connectivity between your system and its attached devices work. In some cases, the amount of software used to make the connection work verges on the astounding when you start to think about the sequence of actions that must occur.

It's not just a matter of creating connectivity between the system and the device either. There are layers to the process. At the lowest level is the operating system, which relies on device drivers to create the connection between devices and the operating system in a standardized manner. The operating system also relies on these device drivers to help in the task of providing access to applications. It's not a good idea to allow a single application to hog a device. Each application must feel as if it has exclusive access to the devices, even when it doesn't.

This chapter won't turn you into a device driver developer or provide you with deep insights into the nature of specific device driver implementations. Instead, you gain a knowledge of how device drivers work in general and what you can do when the device you need to access doesn't quite work as expected. The important thing to keep in mind is that you must have the required physical connectivity and software support in order to make devices on your system accessible and to allow them to work as anticipated.

> **NOTE**
>
> Even though this chapter relies heavily on Windows 7 screenshots, the same principles apply to nearly any operating system you want to install on your system. Every operating system requires some method of layering access between applications and the underlying devices so that there are no device collisions and no one application can hog the device.

Understanding the Operating System to Device Connection

Before you can do anything else, you must have connection between the device and the system. The connection is physical on some level. Even when using wireless devices, some sort of physical adapter sends out radio transmissions to the device and receives radio transmissions from the device. Yes, the intermediate connection is wireless, but the endpoints are wired. There is always a physical connection and a piece of software that accesses that connection. The following sections describe the operating system to device connection in a general way that can apply to any operating system you might use.

Considering the Use of Low-Level Resources

A device is always an extension of the base machine. In order to access any device, the system outputs command to one or more input/output ports. Every device must have unique ports to use in order to communicate properly. For example, Figure 14-1 shows the ports used for a display adapter on a Windows 7 machine. In order to access the device, the address range on the device must match the configured address range for the operating system. Fortunately,

FIGURE 14-1 This display adapter uses two ports with specific address ranges.

modern devices can usually negotiate a usable address with the operating system when you install the device driver, so you don't have to worry about the address unless there is a conflict.

After it issues a command, the operating system can use one or more memory ranges to exchange data with the device. Precisely how this activity occurs depends on the device. A display adapter has a huge amount of data to exchange, so it usually has a number of memory ranges to use. Some devices, such as the COM port, are incredibly simple, so they don't need to use a memory range at all, as shown in Figure 14-2.

The operating system also requires some means of getting the device's attention. The interrupt request (IRQ) setting determines which address lets the operating system interrupt the device and give it a task to perform. As with every other device setting, the operating system and the device must negotiate a unique address for the IRQ. Figure 14-2 shows the IRQ setting for COM1. Unlike other device settings, a device will have only one IRQ address assigned to it. If a physical board contains more than one device, each device on the board will have an IRQ address assigned to it.

An older technology called direct memory access (DMA) also appears as one of the settings for most operating systems. Most devices today don't use

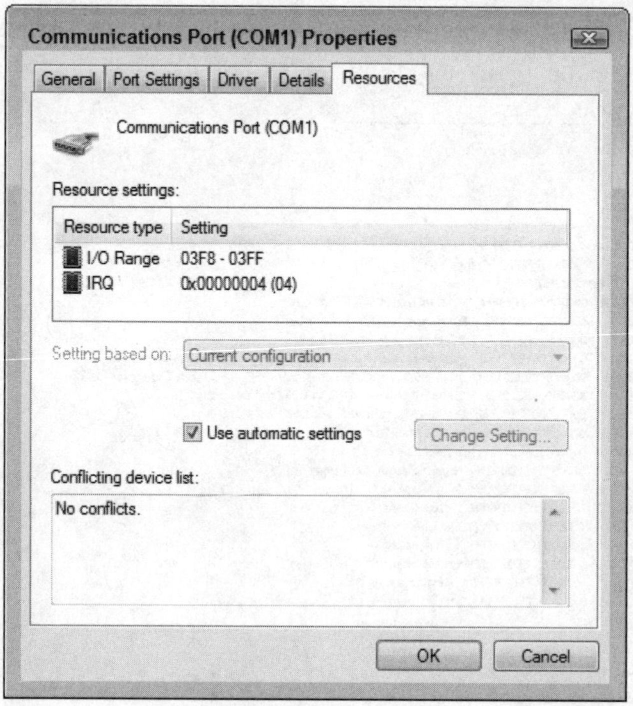

Figure 14-2 Not all devices require use of a memory range.

DMA. However, you may have an older device that requires DMA for high-speed data transfers. As with every other setting, the DMA setting is unique for each device that requires one. Except for the DMA controller, the example system lacks any sort of DMA device and it's unlikely that you'll encounter one either.

Avoiding Device Conflicts

In most cases, the best option for avoiding device conflicts is to allow the operating system to negotiate with the devices through the device driver. The device driver knows which addresses that the device can use. Over the years, vendors have come up with all sorts of ways to create unique port, memory, IRQ, and DMA settings for devices. However, there may be a situation where the device driver and operating system fail to come up with a good solution (a situation that has become incredibly rare). When this problem occurs, you might have to step in and manually provide an address using whatever means the operating system provides for performing the task. For example, when working with Windows, you can view the resources by connection type to determine where an opening might occur as shown in Figure 14-3.

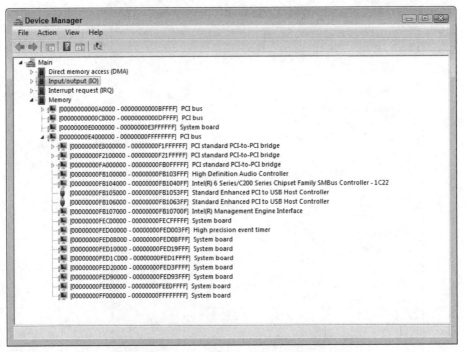

FIGURE 14-3 Every device must have a unique set of addresses to use.

> **TIP**
>
> *Even though you may never need to change the device settings on your system, knowing where to find this information is important. You may need the information when trying to solve a device issue. In some cases, a support person may ask you to make manual changes to a device setup, just to see if a problem goes away.*

Circumventing Application Conflicts

At this point, you know that each device has a single set of addresses. If applications could access these addresses directly, the device would be attached to that application. No other application could access the device, which would cause considerable problems. The operating system maintains tight control over the devices by keeping hold of these addresses. When an application needs access to the device, it does so through an interface that the operating system supplies. Therefore, the first line of defense against application conflicts comes with the operating system.

The second line of defense comes from the application itself. Ill-behaved applications, those that attempt to access devices directly, are somewhat rare today. In fact, you'd need to write special code just to make the attempt because modern programming languages work with the operating system to prevent direct access. In addition, the application would need to operate at a higher privilege level to gain access. Still, it's possible that someone could create an application that breaks all the rules and you need to be aware of the potential for problems when an application does so.

It's possible that an errant application could so overwhelm the operating system with requests that the device could become effectively locked. This is especially true of some devices such as a sound board. If you've ever encountered a situation where nothing you can do will keep the sound system from replaying sounds or acting oddly in some other way, you've seen the effects of an application lockup. Buying high-quality, well-designed applications will solve this problem. Chapter 15 goes into techniques for buying great applications in more detail.

Relying on Operating System Drivers

When you initially install an operating system, all of the devices attached to your system will use operating system–specific device drivers unless you made a special effort to install a vendor device driver during the installation process or there is no operating system device driver available. The operating system device drivers are sometimes supplied by the device vendor, but could also be supplied by the operating system vendor or a third party. The following sections describe operating system device drivers in more detail.

Defining the Advantages of Automatic Access

The main advantage of operating system device drivers is automatic access to your device. For example, when you plug a new device into the Universal Serial Bus (USB) port, the operating system automatically detects the device, looks for a device driver, and installs that driver on the fly. All you see is a device that appears to function magically almost the second that you plug it in. However, behind the scenes, the operating system is doing quite a bit of work on your behalf to ensure the device works.

Considering the Missing Device

In some cases, especially when working with less common devices, the operating system won't find a driver to support the device. In this case, the operating system may go online (assuming you have an Internet connection) and look for a device driver for you. However, even in looking online, the operating system may fail to find a device driver that will work. In this case, the device is inaccessible.

Your first clue that the device isn't working is that you can't access its functionality. Of course, you don't know why the device isn't functioning and assuming that it's the result of not having a device driver is a bad idea. The "Overcoming Driver-Specific Problems" section of this chapter delves deeper into diagnosing device problems, but the essential consideration now is that the operating system will indicate a missing driver in some way. For example, when working with Windows, you see the device listed using a generic name with an exclamation mark (bang) next to it. The device may not appear in the right category and is unlikely to have the right name, but you at least know that the device isn't working as it should. A clear Device Manager display (like the one shown in Figure 14-4) tells you that all the devices on a system are working correctly and have a good device driver installed.

Obtaining and Using Device Vendor Drivers

Many devices work just fine using a generic operating system device driver. For example, the COM port on your system doesn't really need anything special to make it work. You can often get by without a special device driver for your mouse and keyboard as well. However, when it comes to other devices, such as a display adapter, using a vendor-specific device driver has significant advantages. The following sections describe these advantages and provide you with some ideas on how to make best use of vendor-specific device drivers.

Chapter 14: Accessing the Devices

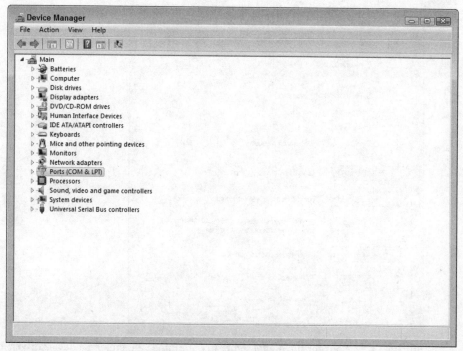

FIGURE 14-4 Look for a clear display when checking for missing devices.

Considering the Advantages of Vendor Drivers

Every device driver on your system has a vendor name or other identifier attached to it. For example, if you have a device that uses a generic Windows device driver attached to it, the digital signer for that driver will appear as either Microsoft or Microsoft Windows as shown in Figure 14-5. No matter which operating system you use, the operating system provides some method of finding out who produced the device driver that creates the connection between the device and the operating system.

When you see an operating system signer for a device driver, you know that the device driver is generally generic in nature. Depending on the device in question, upgrading to a vendor-specific device driver could have the following advantages:

- **Access to additional features:** Generic device drivers can support a number of devices with the same general categories of features. A vendor-specific device driver knows about all the special features that a device provides.
- **Greater speed:** With some devices, the speed difference between a generic device driver and the vendor-specific device driver is quite evident. The vendor-specific device driver will know about parallel processing and other features that the device provides. The difference is exceptionally apparent when working with some devices such as display adapters.

FIGURE 14-5 Look for the digital signer to determine who created a device driver.

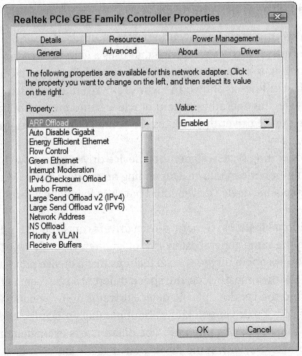

FIGURE 14-6 Vendor device drivers often provide additional configuration features.

Chapter 14: Accessing the Devices

- Higher reliability: A generic driver can glitch when a mismatch between the device and the driver occurs. Not every vendor follows every rule for creating devices precisely. A vendor-specific device driver knows when the vendor has taken a different path.
- Lower power consumption and noise: Many devices have fans and other features that make noise. In addition, these devices can support modes where they power down unneeded features. The combination of shutting off fans and going into power-saving mode results in lower power consumption, but only vendor-specific driver will know about this functionality.
- Additional device driver configuration: Vendor-specific device drivers can provide additional configuration options that you won't find with a generic driver. These configuration features can include advanced settings and how you want to manage power or other resources. Figure 14-6 shows a typical example of a device driver that provides additional functionality.

Keeping Vendor Drivers Updated

Some vendors do update their drivers automatically. For example, when you install a display adapter and its associated software, you generally receive automatic updates as part of that software. Figure 14-7 shows the setup for a NVIDIA GeForce display adapter. In this case, you can even choose specifically when to perform updates.

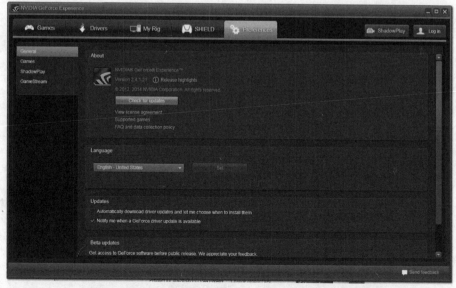

FIGURE 14-7 Some devices come with complete support for driver updates.

Unfortunately, many devices don't provide automatic updates, update checks, or any other sort of notification. In some cases, you must manually download the driver from the vendor site to obtain an upgrade. The additional work is necessary to keep your system running at peak efficiency and ensure you don't have problems with hackers. The best way to handle this situation is to check for driver updates on a regular basis by going to the vendor sites.

Updating vendor drivers without thinking things through is an almost certain way to create an unbootable system. For this reason, you want to be notified about updates or check for them manually, but you only want to install the updates at the right time. Because device drivers are prone to making your system unbootable, make sure you always have a fresh backup of your data files before you install the device driver update. The following sections provide some additional considerations for device driver updates.

Synching Device Drivers to Operating System Release

When working with the generic device drivers provided with an operating system, the operating system vendor usually ensures that your system meets the requirements for using the device driver before installing it. Vendor-supplied device driver updates may not come with the same guarantee. However, vendor-supplied device driver updates usually come with some sort of documentation telling you the requirements for the update. Make sure you check on the requirements before you perform the update (or face an unbootable system). For example, if the device driver says that it supports Service Pack 2 of an operating system and you only have Service Pack 1 installed, make sure you perform the operating system update before you install the device driver.

> **NOTE**
>
> *It's not unusual to find discussions online where someone has tried an unconventional device driver configuration or performed an update outside the normal requirements. Given that someone else has tried it, you might be tempted to try it on your system as well. If the other person's system is precisely like your system and you feel comfortable enough with their procedure, you can certainly try the device driver. However, always assume that the new device driver won't work, removing it isn't an option, and that you'll have to install your operating system and applications from scratch before you even attempt to perform the task. In other words, assume the worst-case scenario for your actions and be happy if the process does indeed work as hoped.*

Chapter 14: Accessing the Devices

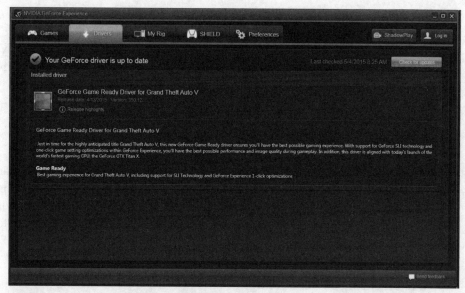

FIGURE 14-8 Some vendors will release special versions of a driver for specific applications.

Understanding Application Requirements

Even when a device driver will work with the operating system, it isn't certain that it will work with your applications. For example, a display adapter driver update may make it possible to perform advanced tasks with the latest version of a game. However, using the device driver update with the older version of that game might not work at all. Sometimes a vendor will even release a special version of a device driver for a specific application, such as the device driver shown in Figure 14-8.

Application oddities can result immediately after installing a device driver. For example, on one system I tested, updating the display adapter driver would render Firefox less than usable. The solution was to perform the update with Firefox closed, reboot the system, and then restart Firefox. You won't find this requirement documented anywhere. However, it's an example of a subtle application-specific problem that can occur. Fortunately, in this case, there is a relatively easy workaround.

> **WARNING**
>
> *Installing device driver updates that are incompatible with specific applications can cause your system to freeze when you start the application or incur data loss. However, application incompatibilities are seldom fatal to the system as a whole. If you encounter such a problem,*

you can look for a fix from the application vendor. When the application vendor lacks a fix for the problem, uninstalling the new device driver and reinstalling the old device driver may be your only solution to fix the problem. The bottom line is that it's always better to check compatibility before you do anything, rather than to have to fix problems later.

Installing the Device Drivers

Installing a device driver is a process. Even if you have driver-support software, such as the NVIDIA product shown in Figures 14-7 and 14-8, you need to ensure that the driver installation will be a success before you do anything. The following sections provide you with a path that you should follow when installing device drivers, especially those from vendors.

Documenting Driver Settings

Figure 14-6 shows a device driver with additional settings. Any update you perform is likely to change any customization you perform, which means that your system might suddenly act sluggish, have application compatibility issues, or encounter other problems. Of course, just trying to remember what you did to get rid of those problems is going to be a nightmare. It's always best to document any settings changes immediately after you make them. However, even if you fail to keep a log of existing settings, ensuring that you document any custom driver settings before you install an update is a great idea.

TIP

Some driver-support software provides a settings backup feature. All you need to do is tell the software to perform a backup and a file with the settings appears on your hard drive. Using the driver-support software option is quick, easy, and usually less risky than recording the settings by hand. However, make sure you check online to determine whether anyone has had a problem with the driver-support software missing important settings (resulting in problems after an update).

Performing a System Backup

Losing a setup when you first build a system is hard; losing a setup and associated data after you've used a system for a while can be devastating. Prior to performing any major system undertaking, always create a backup. Yes, there is the time factor involved. Many people want to get the update installed and immediately start back to work. However, instead of thinking about the time spent doing the backup, think of the time saved by not having to restore your data later (possibly typing it in by hand).

Removing Existing Drivers

The driver update process often involves replacing one driver with another. In fact, many operating systems offer an option for simply updating the driver. This approach works in most cases because the operating system is performing a lot of work in the background for you. If this support doesn't exist, then you might have to uninstall the old driver first and install the new driver afterward. Depending on the operating system you use, uninstalling the driver may not amount to much more than running an appropriate application or clicking a button like the one shown in Figure 14-9.

After you uninstall the driver, you may have to perform a system reboot to remove the driver software from memory. During the reboot process, the operating system may replace the custom driver you removed with a generic driver. You later replace the generic driver with a new custom driver, as if you were installing the operating system from scratch.

FIGURE 14-9 Uninstalling the device driver is as simple as clicking Uninstall.

> **NOTE**
>
> *Always read the vendor driver documentation to ensure you use the correct process for removing and installing device drivers. Removing an old driver could also remove settings that you could otherwise keep—costing you additional reconfiguration time. Remove or replace a driver according to vendor instructions.*

Performing Required System Reboots

Device drivers operate at a relatively high privilege level. In addition, the operating system loads the device drivers during the boot process and the device drivers remain in memory until the operating system shuts down. As a result, removing a device driver can prove extraordinarily difficult without the appropriate software (such as a vendor-specific application of the sort shown in Figures 14-7 and 14-8) or by rebooting the system. The following list provides an overview of the times when the system could require a reboot in order to complete a device driver removal, update, or installation:

- The operating system displays a message at the end of the device driver removal, update, or installation.
- The operating system displays error messages when you try to install a new device driver.
- The device driver action keeps failing—it doesn't appear to complete.
- Applications act oddly after you perform the device driver activity.
- You suddenly lose access to device features.
- The operating system seems unstable or performs odd actions (such as requiring extraordinarily long times to write data to disk).

Installing the New Driver

You install new device drivers after installing a new device, upgrading an existing device, as part of replacing a generic device driver with a vendor-specific device driver, or as part of updating an existing driver. For example, when you click the Update Driver button shown in Figure 14-9, what you actually do is

1. Uninstall the old device driver.
2. Remove the device driver from memory.
3. Install a new device driver.
4. Loads the new device driver into memory.

All of the details of performing these tasks are hidden in the background. However, it's important to know that the activities are taking place because you never know when the operating system will suddenly decide to perform a task incorrectly. Knowing how the task works can help you troubleshoot the problem and recover later.

Beside updates, device driver installations can occur in a number of ways. The methods that are available to you depend on your operating system and the device in question:

- Operating system support: Sometimes when you look for an update, it's already available on your hard disk as part of an operating system service pack. Normally the update installs automatically, but you need to be aware that the operating system may fail to make the required changes.
- Inserting a disk: When you buy a new device, it usually comes with a disk that contains the device driver. Inserting the disk starts an automatic process that installs the device driver for you (or at least provides a menu that tells you that the installation option is available).

> ### WARNING
> *The device drivers on a disk supplied with a device are normally outdated. They work long enough for you to get your system up and running. However, once your system is stable, you should find an updated version of the device driver to ensure you gain access to the latest features and security fixes.*

- Automatic download: The operating system, the device, or some other software may look online for you and automatically download the required software.
- Manual download: You go to the vendor site, look up your device, and match the downloadable version of the device driver with the version already installed on your system. When the driver on the vendor site is newer, you download the new driver file, open it, and then follow the instructions for installing it.

Overcoming Driver-Specific Problems

Device drivers are major part of your system. A sick device driver can cause all sorts of woes, some of them quite hard to figure out. The fact that a device driver runs at a high privilege level and can cause so much damage to the operating system, applications, data, and sometimes even hardware is the reason that operating system vendors validate device drivers. In order to install the device driver, it must be signed off by the operating system vendor. However, despite the best efforts of everyone involved, sick device drivers still end up on systems and cause problems. The following sections help you diagnose and potentially fix the most common device driver problems.

Ensuring Device Connectivity

It may seem odd, but devices become detached. Many devices have clips or screws to keep them firmly attached to the system, but even so, vibration and user activities can cause the device to become detached from the system. It may even look like the device is attached, but a slight wobble has introduced just enough of a connectivity problem to cause the device to malfunction. Before you do anything else, unplug the device and then plug it back into the same connector. You may be surprised to learn that the device simply had a connectivity issue.

> **WARNING**
>
> *Some devices, such as those connected to a USB port, are designed to allow you to plug and unplug the device while the system is powered up. However, other devices will potentially spark and could possibly fail when you connect or disconnect them while the system is running. When in doubt, power the system down before you unplug the device and then plug it back into the system.*

Disabling the Driver Temporarily

Assuming a device driver is bad usually isn't a good idea. You need some type of verification that the device driver is the source of problems on your system. When working with noncritical devices, such as a sound board or COM port, you can usually disable the driver temporarily to see if the problem disappears. If it doesn't, then you know that the device driver is probably clean.

> **WARNING**
>
> *Never disable a critical device. For example, disabling your display adapter is a really bad idea because then you can't see what's going on with the system. Rebooting the system will make things worse because now you're not logged in and still can't see what's going on. Likewise, the keyboard, mouse, processor, motherboard, and other essential devices must remain functional while you diagnose a problem.*

Disabling a device driver doesn't remove it from the system. All that it does is make the device driver temporarily unavailable. The device driver goes to sleep until you enable it again. Because the device driver isn't active, you can see the effect on the system and determine whether the device driver was at fault.

Chapter 14: Accessing the Devices

Unfortunately, the device driver is also loaded into memory. You may have to disable the device driver and reboot the system to determine the true effect on system functionality. After the system reboots, you know that the device driver is no longer in memory and can truly see the result of removing it (hopefully, the change isn't of the sort that will cause you to rip your hair out).

After you check the device driver, enable it again. As with disabling the device driver, you might find that the device still isn't available. Before you make any other changes to the operating system configuration, make sure you reboot the system to reload the device driver into memory.

Verifying Driver Settings

As mentioned in the "Considering the Advantages of Vendor Drivers" section of this chapter, vendor drivers often include extra features that you can configure to obtain maximum performance. Generic drivers can also include these sorts of settings as shown in Figure 14-10. Any update you perform is likely to cause the driver settings to return to a default state, usually factory settings, which may not work with your system. The odd thing is that installing other sorts of software can also cause the driver settings to change. It's important to maintain a list of

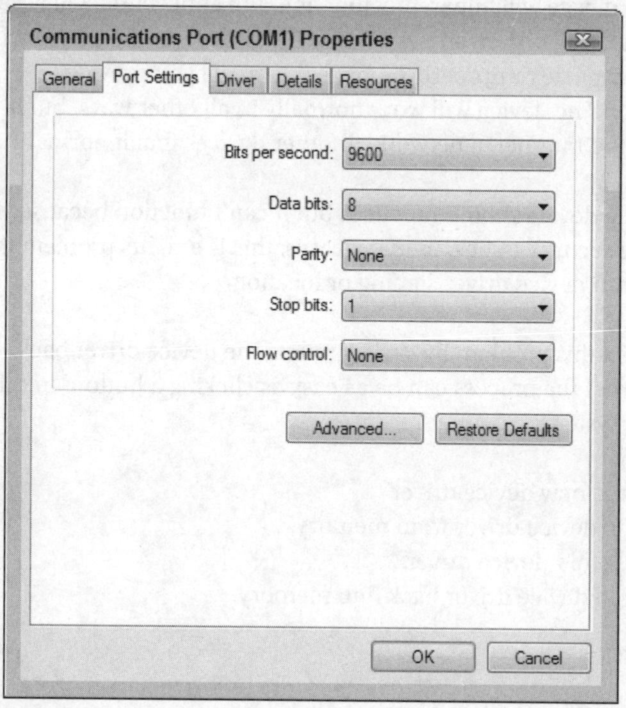

FIGURE 14-10 Some generic device drivers also have special settings.

workable settings and then check those settings when the device driver appears to have problems.

If you find that some of the device driver settings have changed, you need to return them to the workable state for your system. However, because the device driver is in memory, you may need to reboot the system before the changes will take effect. If you don't see a change after waiting for a minute or two (the change can take time to implement), then a reboot is probably necessary to invoke the new settings.

Considering the Device Driver Mismatch

Mismatches are a problem that can be hard to diagnose (or incredibly obvious). It's important to understand the mismatches occur at several levels:

- Device: In which case, the device is inaccessible.
- Operating system: The operating system can fail to recognize the device or become unstable.
- Application: An application will freeze, act oddly, refuse to interact with the device, or misbehave in some other way.
- Data: An application will claim that it can't interact with the data in some way, even if you have worked with the data in the past.
- Media: The device will apparently function, the application will access it, and the operating system shows no stability issues, yet the device will refuse to recognize or properly manage the media you provide.
- Connectivity: The device will work normally in all other ways, but it won't connect to the Internet, a network, or other devices that it normally works with.
- Security: The device should function, but it can't function because you keep seeing security issues. In many cases, this is a rights problem that has to do with device driver signing or location.

It's possible to fix this problem by simply rolling the device driver back. Figure 14-9 shows that the process can be as easy as clicking a button. In this case, the operating system:

1. Uninstalls the new device driver.
2. Removes the device driver from memory.
3. Reinstalls the old device driver.
4. Loads the old device driver back into memory.

15

Choosing Applications

A functional system requires more than an operating system. Up to this point, you have built and configured what amounts to a blank canvas. The canvas has a size, shape, and tone that are defined by the components and operating system you chose for your system. If you were to show your system off to your friends, their reaction might be a little bland, depending on their understanding of the nuances of hardware, operating system, and configuration. However, show the same system to your friends with a game running on it and it suddenly becomes quite exciting. Depending on how well the system interacts with the applications, the excitement can reach a fever pitch. The applications present the image of your system as an extension of yourself. Choices you make in the selection, configuration, and use of applications are the paints that draw the picture of your system on the canvas you've created to date.

Just as an artist carefully chooses a subject, technique, and medium for an image, you must exercise care in working with applications. In fact, the application process follows these basic steps.

1. Determine which applications will actually work with your system.
2. Ensure the application provides the features and functionality you want.
3. Install the application on the system.
4. Define and correct any application issues that prevent you from having a great experience with it.

The sections of this chapter provide you with details that you might not normally have considered as part of an application purchase and installation. The use of a custom setup means that you gain speed, reliability, and security benefits that people who obtain off-the-shelf components could never expect to get. However, custom setups also come with some challenges when it comes to application selection, installation, configuration, and debugging.

> **NOTE**
>
> *This chapter discusses applications in a non-platform-specific manner. You can use the advice presented with any operating system of your choice. In some cases, a specific operating system version may require additional attention to detail that this chapter doesn't provide. Make sure you always read the vendor documentation as part of getting ready to make application purchases. In addition, it doesn't hurt to perform a little research online regarding the options available to you. Your peers are always finding ways to tweak a setup to accommodate seemingly impossible application setups and configurations—making use of their experiences is always a good idea.*

Matching Applications to Your System

No matter how much you desire to play that Mac game on your Windows system it probably won't work—at least, it won't work well. People are constantly shoehorning applications onto systems that aren't designed to accept them. Yes, many of them finally get the application to work, but the quality of the experience suffers. What finally happens is that the user spends considerable time propping the application up and making it work in a situation it was never intended to address in the first place. The application doesn't work well, the user is frustrated, and no one is happy. The point of building a custom computer and creating a unique configuration for it is to allow you to have a maximum of fun with a minimum of pain. Of course, this means getting applications that really are a good fit for your system. The following sections describe some of the issues you need to consider.

Verifying the Hardware Requirements

Most applications tell you the minimum requirements required to make them work. If you find an application that doesn't provide this information, make sure you research the information before you buy it. You can't be sure an application will work if you don't know what it needs in the way of resources to work properly. However, the information you obtain can be misleading for a number of reasons.

It's important to understand that the minimum requirements usually represent an unencumbered system. If you run other applications at the same time that you run the target application, you must also consider the needs of those other applications. When you overload memory, processor, or hard drive, you get less than useful results, even when the system would normally run the application without any problem. When running the application with other applications, consider these requirements:

- It's possible to share the processor, but you still need more processing cycles when running multiple applications.
- You can't share hard drive space. It's a good idea to have at least twice the minimum space available on your hard drive before you begin an installation. More is always better when dealing with hard drive requirements.
- Real memory is expandable with virtual memory, but you must have enough memory for all the applications you want to run.
- Virtual memory is not the same as real memory. You must have enough real memory to allow the application to load its active part. At least half the memory requirement for an application should be in real memory.
- Peripheral devices are completely shareable, so the mouse you use with one application is completely available for any other application. You normally need just one of each peripheral device to satisfy an application requirement.

Vendors want you to buy their application. When a vendor lists a minimum requirement, it means the minimum required to run the application at all. The minimum usually doesn't provide good reliability, high speed, or even great security. If you want to see your application at its best, always provide more than the minimum requirement.

Validating the Platform Requirements

Applications generally tell you what platform they need to run. In some cases, the application will tell you about specific platform requirements. For example, you might need to have a particular service pack or platform-specific feature installed to make the application work. The application could work on a system that doesn't completely meet the requirements, but application functionality, reliability, or security may suffer as a result. The best application performance comes from a platform that completely meets the application needs.

Many people rely on emulators to create the required platform environment. Even though emulators can make the application run, using an emulator doesn't generally provide great results. An emulator will generally work fine with business applications. You may even get a game to work properly using an emulator (although, you shouldn't count on it with complex or high-end games). However, as the application becomes more complex, the chances of getting it to work with an emulator become increasingly small. In general, avoid using an emulator whenever possible to stand in for a real operating system.

Checking Version Information

Different versions of an application have different needs. You want the newest version of an application that fits your hardware and platform specifications. What this means is that you may not always want the newest available version of an application, but simply the newest version for your setup. It's possible that a vendor will have multiple versions of an application available at one time to ensure that everyone who wants to use the application can do so.

> **TIP**
>
> *Once you know which version of an application works with your setup, make sure you verify the version every time you research information about the application. Some reviews, articles, and online discussions relate to a specific version that may not reflect the version that you plan to get. A common source of misinformation occurs when the version of the information provided by a source doesn't match the version of the application installed on the system. Misinformation causes users all sorts of problems (such as misconfiguration) and applying misinformation could keep the application from working at all.*

Considering Connectivity Requirements

At one time, you could normally get by running an application without any sort of connectivity. However, newer applications can require several forms of connectivity and you need to ensure your setup meets these needs. When a system doesn't meet the connectivity requirements, it may operate at reduced functionality, block certain features, or not work at all. The issue of connectivity is especially important when working with games because many modern games perform a license check every time you run them (to ensure you have a paid, valid copy of the game).

It's likely that you do have an Internet connection. However, the speed of that connection is also important when an application relies on the connection to perform basic or essential data exchanges. When the connection is used for application updates or simply for a license check, the speed of the connection is less important, but you need to know in advance how the application uses the connection or you might have a nasty surprise when the application doesn't work as anticipated.

Interestingly enough, some applications also require a local area network (LAN) connection. These applications often see use in collaborative environments. Yes, you could use them alone, but the developer has the expectation that you won't use them in this fashion. In some cases, you can configure the application to get around this particular need. Before you make the application purchase:

Chapter 15: Choosing Applications

- Research all of the connectivity the application requires.
- Determine whether the connections are required or optional.
- Verify the connection type.
- Consider how the application uses the connection.
- Obtain the minimum connection requirements, such as speed and reliability, and match them to application connection usage.

Obtaining Application Reviews

A review is an opinion. No matter who writes the review—whether the person is famous or not—every review you read is an opinion. The opinion may not match your opinion of the application or reflect how you feel it works. Even the reviews in this book are all opinions, they happen to be mine. Whether you agree with a reviewer in general depends on how the reviewer and you view products in general. You may find that you disagree with your favorite reviewer on occasion. The view of reviews as opinion all leads up to the fact that you can't base an opinion about the usefulness of an application for your specific setup on a single review—you need as many reviews as possible to determine whether an application meets your needs and delivers what it promises.

Reviews come from many different sources. Professional reviews found in magazines tend to provide a more balanced review because editors and other parties help ensure that the review you see is fair to an extent. However, editors are also cognizant of the fact that vendors pay the bills, so you won't see the worst potential issues of an application without reading personal reviews on sites such as Amazon. Of course, the reviewers on sites such as Amazon don't really take responsibility for what they say, which is why blogs also provide a good source of reviews. A review that demonstrates the product is always better than one that simply expresses an opinion because the demonstration helps you see product functionality that the vendor may not make apparent. Using as many different sources as possible when choosing an application to install on your system is important.

> **WARNING**
>
> *Some people think that reviewers are innately fair and unbiased, but this viewpoint is far from the truth. In fact, you find seeded reviews (those provided by a competitor to make an application look bad) all the time. No matter who writes a review, the review is always biased because the writer would find it impossible to present a view of an application that doesn't rely on personal experience. In addition, despite the best efforts of some reviewers, most reviews are unfair because a testing technique that works with one application may not work with another. About the best*

that you can hope to achieve is a balanced review where the reviewer provides a demonstration that describes the features, good points, bad points, and application statistics (including things like vendor contact information and application price).

Installing the Application

At some point, you've considered every aspect of the application and you feel certain that it will work on your system. In addition, you know enough about the application to feel it will meet your needs and provide enough of what the vendor has promised to work fast, reliably, and securely. It's time to perform an installation. Unfortunately, there isn't any sort of template for installing every application on the planet. The following sections provide you with an overview of the general considerations for installing applications. However, you also need to reference the vendor documentation to obtain the full installation instructions.

> **TIP**
>
> *Reading the documentation is your first defense against a bad installation. Looking over the README and other files that come with the application installation files is your second defense. Referring to the installation experiences of other people is your third line of defense. In short, reading before doing is the only way to ensure that you'll receive the sort of installation that you really want from the application.*

Avoiding Installation Complexities

The biggest problem that most people encounter is a lack of organization. Make sure you have everything needed to perform an installation before you begin. Fortunately, modern applications make getting ready a lot easier. At one time, you needed to provide all sorts of really strange information, such as the setup and configuration of system devices. Your operating system provides this sort of information now. However, modern applications still need you to provide information such as

- Installation location
- Licensing information
- Personal information
- Location of online connections
- Special equipment access

The best installation to use is the express or typical option. In fact, you should use this option whenever you can because the installation program will look up many bits of information automatically. Even if you have to perform some configuration after the application is installed, you still save time and headaches by using an express or typical installation. The more complex the application, the better it is to use the simplest possible installation option.

Some applications provide several levels of custom installation. Use the least complex of these options whenever possible. The fewer questions you answer, the less likely it is that you'll answer one of them incorrectly (often, despite knowing the right answer).

When you must perform a fully custom installation, try going through all of the installation steps first and cancel the actual installation before it actually starts unless you're absolutely certain that you answered all questions correctly. If you have any doubt about the answers to custom installation questions, make sure you research the answers to those questions before you start the installation again.

> **TIP**
>
> *Some application installations are complex enough that you can find installation walkthroughs online. Make sure you verify that the demonstrator is using the same version of the application you are and then go through the walkthrough. Note any areas where you might encounter problems and be sure to research those areas before you proceed. Walkthroughs often describe the ways to get answers to your questions—saving you a lot of time as a result.*

Using an Installation Program

You normally use an installation program when an application installation is fairly complex. The installation program ensures that the application appears in the right location on the hard drive and that the operation system knows about the application. In many cases, an installation program will perform a number of additional steps for you:

- Verify the system can actually support the application
- Ensure you have any required additional information
- Check for potential problems such as a lack of connectivity
- Determine whether you need any system or device updates
- Install any required support files that aren't actually part of the application

Installation programs often have names such as Install or Setup. Sometimes they have complex names that define the platform they support. Whatever the name of the installation program, you usually start it by typing its name at the command or terminal prompt, or by double-clicking its entry in a folder. All you do once the installation program is started is follow the prompts on screen.

Unpacking the Application

Simpler and open-source applications are often self-contained. You receive some sort of archive file, such as a .ZIP or .TAR file that contains everything needed by the application. All you have to do to use the application is to unpack it in an appropriate folder on your system. The advantage of this approach is that it's incredibly simple and the least likely way to install an application that will cause any sort of a problem. However, this approach also has some potential issues you need to consider:

- There is no way to know whether the application will actually run because there is no installation program to check system compatibility.
- The operating system is unaware of the application, so you don't get any operating system support, such as automatic file extension recognition.
- You must install the application in the correct folder. Failure to choose the correct folder can cause the application to fail due to a lack of resource access or security issues.
- Creating any system-specific functionality, such as application access through operating system menus, is a manual process.

> **NOTE**
>
> When you unpack an application, it usually contains a README or other documentation file. In fact, you may find several documentation files in the archive. Make sure you actually read and understand the documentation before you begin using the application because the application may have configuration or other requirements that you must address before you can use the application.

Working with Web-Based Applications

Many applications today are web-based applications, even when they look like they're a common desktop application. In many cases, the application downloads a shell to your system that makes the application look more like a desktop application. The shell may also store information locally and make it possible to perform other tasks. The differentiating factor for most web-based applications is that you start by accessing them on a site and that the installation process normally occurs through a browser interface.

When working with web-based applications, you need to consider the site you're working with and connectivity becomes a significantly greater problem. It's also important to place a much higher emphasis on security due to the proliferation of web-based viruses, Trojans, and hijacking. The best approach is to do your research before you install a web-based application and then keep your eyes open for news stories regarding the product.

Chapter 15: Choosing Applications

> **TIP**
>
> *When working with web-based applications, keep security on your system as a whole quite high, including the browser. Grant the site only the security rights it requires so that you can run the application. Keeping security tight reduces the number of potential holes a hacker has to attack you. However, never assume your security is airtight. Always assume that someone can get into your system if they're truly motivated. Monitoring your system can help you detect security issues before they become a problem:*

- Watch for sudden losses of system responsiveness and applications that no longer run as quickly as they should.
- Check for sudden decreases in hard drive space.
- Look for directories that appear and reappear even though you haven't installed anything new or reconfigured your application.
- Verify the source of sudden increases in application behavioral problems, such as freezes or data loss.

Correcting Application Installation Problems

Most people experience installation problems at some point. There simply isn't a way around the problem because systems use a variety of components and installation programs don't always work as advertised. Mismatches, missing support software, and even glitches in your hardware can ruin a perfectly good installation. The following sections provide some insights into the ways in which you can deal with application installation problems.

Dealing with Mismatches

There are times when you have done everything possible to ensure the installation will go as planned and it still doesn't succeed due to some sort of mismatch. Even though it appears that your system is composed of a wealth of individual parts, those parts have to work together in order to create an environment where your application can do its job. With this in mind, consider these potential sources of mismatch:

- Device drivers
- Operating system configuration
- Patches and service packs
- Unsupported hardware
- Other software

It's important to find the source of the mismatch. In many cases, this means doing some research to find potential candidates. Perhaps other people have had the same problem and can provide you with insights as to where to look. You can also try removing one (and only one) potential source of problems at a time until you find the specific source that causes the problem. Until you know the specific source of problems, the installation will continue to fail or create an application installation that has speed, reliability, or security problems.

Some issues are relatively easy to fix. For example, you can generally find an update for a device driver that isn't working correctly. An operating system may provide alternative settings to help create an environment where the installation will work properly.

A few issues are quite difficult to fix. For example, when you need a particular piece of unsupported hardware, you might find it hard or impossible to replace it. Perhaps there isn't an alternative device available. At some point, you need to decide which is more important, the source of the conflict or the application you want to install. In some cases, finding an alternative application is cheaper and faster than trying to fix the problems of an errant application.

Finding Missing Components

You must have all the components required to run an application. Advanced installation programs check for these components and let you know when they're missing. Even less advanced installation programs will provide you with the name of a missing file. Researching the filename often tells you what you need to install in order to make the application work. However, sometimes you find that there isn't any significant help available. Researching the application can sometimes yield results—relying on peer help can help in others.

> **TIP**
>
> *There have been more than a few times when an application failed to work for lack of components that the components have actually appeared in subdirectories of the application installation package. The installation program never attempted to install the components or even allude to their presence. Only by actually looking at the installation package content (including those all-important README files) was it discovered that the needed components were indeed available and ready for installation.*

Getting Peer Help

Peer help can be quite useful, but it can also be quite tricky. Your peers have probably encountered the problem you're having at some point and solved it.

Making use of peer help means that you learn from other people's mistakes and gain their insights. However, peer support can also be fraught with problems. Some people provide less than useful help—some of which could actually damage your system. You might be the first one to encounter a problem, but that won't stop people from guessing at solutions that may not work (in fact, chances are that they won't). When relying on peer support, make sure you keep in mind that the help doesn't come with any sort of guarantee and if you damage your system, you're on your own in fixing it. Here are some suggestions for making it more likely that you receive good peer support:

- Verify that the site you want to use to post the message actually works with the product you're researching.
- Look at other queries to see how often the group actually provides useful information to other people.
- Provide complete information in your question, including your platform and system setup particulars as needed.
- Post the question only once and only in the most appropriate area.
- Pay attention to the questions you post and make you provide prompt responses to any queries people make in order to answer your question.
- Treat others with respect, even if the advice they provide doesn't help in your particular situation.

Resetting an Installation

It happens to everyone. Sometimes you simply get a bad installation and there is no explanation for it. You couldn't repeat the sequence of events that led to the bad installation if you tried because generally you aren't to blame for the installation problem. In some cases, the hardware glitches from a power fluctuation, heating effect, or even solar radiation—causing the software to install incorrectly. Lost bits during a download or timing problems can occur as well. In fact, there are myriad reasons that an installation can fail only once and trying to find out the specific reason is like looking for a needle in a haystack.

Fortunately, the solution to the installation that fails for no apparent reason is quite simple. All you need to do is redo the installation. Unfortunately, if you just run the installation program, the probability is that the installation could fail again. Following these steps improve your chances of getting a good installation the second time around:

1. Uninstall the application (assuming it provides an uninstaller).
2. Remove the application's installation directory (when there is one) and permanent directory from your hard drive.
3. Remove any configuration files or other application data from your personal directory.

4. Reboot your system to remove any data that might have lodged in memory.
5. Perform the reinstallation.

Removing Errant and Unused Applications

Some applications just won't work out. It doesn't matter how careful you are or how much time you spend in research, some application decisions turn out bad. When you decide that you really don't want an application, don't just leave it there cluttering your system. Here are some things to consider about unused applications:

- Keeping your system clean makes more resources available for applications you do want.
- Removing unused applications reduces the number of security holes on your system and reduces the chance that a hacker will gain access to it.
- Deleting old applications (especially applications that caused problems in the past) reduces the chance of conflicts when you install a new application.

Part V
Performing Maintenance

16
Maintaining the Hardware

Building a custom PC is an expensive process (as you well know by now). It's not only expensive to buy the parts, but you also invest your time in building the PC. Your time is most definitely worth something. The cost of your PC is a lot more than simply the price of the parts, so it's essential to maintain that investment. After all, you want to use your new machine for quite some time to get the maximum benefit from it.

This chapter provides you with the techniques needed to maintain the physical device—the computer hardware. It's important to perform all of the tasks presented in this chapter because they're all important. The buildup of dust and grime on your system doesn't just present an unpleasant appearance—it can also reduce the life of your system. Some dust, for example, contains metallic elements that can actually cause system shorts.

Part of maintaining your equipment is to perform the task regularly. The last section of this chapter provides some guidelines you can use to schedule maintenance. Obviously, you don't want to perform maintenance every day, but you do have to perform it often enough to make the maintenance effective.

> **WARNING**
>
> *For your own safety and the safety of the computer components, make sure the computer is off when you perform maintenance on it. The only time you need to turn the computer on is when you need to perform tasks such as verifying that the fans are still operational. Otherwise, most tasks will work better with the computer off.*

Cleaning the Outside

You should clean the outside of the case and the associated peripheral devices more often than you perform other maintenance because these surfaces take a lot of abuse on any given day. In addition, they're open to any external sources of dust and damage. The following sections describe how to clean the outside of the case without potentially damaging it.

Vacuuming the Louvers and Cables

The louvers allow air inside the computer. As air flows through them, they collect dust—sometimes a great deal of dust in a relatively short timeframe. It doesn't matter how many fans your system possesses if the cool air can't get through from outside the case. So, keeping the louvers clear of dust is essential.

It's possible to use the hose attachment of any standard vacuum to clean the louvers. Simply suck the dust off the louvers. Don't brush the louvers because you may actually push the dust inside.

When cleaning the cables that attach to the back of the computer, carefully suck any dust off. Don't brush the cables because you could damage them. In some cases, it's actually better to use compressed air to blow the dust off the cables, rather than try to suck it off. Remember to clean the inside of any plugs that aren't in use.

> ### *WARNING*
> *Some vacuum cleaners have a relatively large motor that has a significant magnetic field. The magnetic field could cause damage to the magnetic media (such as the hard drive) inside your system. Keep the vacuum motor away from your computer and rely on the hose attachment to vacuum it instead. If you have any doubts about the usability of your vacuum with the computer, buy a computer vacuum to perform the task—it has a small motor that is unlikely to affect the magnetic media.*

Wiping the Case

After you vacuum off the dust, you want to carefully clean the outside of the case. Doing so will help keep the case looking nice, but it also lets you see any damage to the case or the external computer features with greater ease. The following sections help you clean the outside of the case.

Dusting Hard Surfaces
Use a soft cloth, such as a hand towel or an old tee-shirt, to wipe the hard surfaces on the outside of the computer. Never use a wet cloth to perform this

task. Try to clean the outside of the case using just a dry towel first. If absolutely necessary, you can use a small amount of spray cleaner to remove stubborn dirt. Try spraying the cloth first to see if a dampened cloth will do the job. It's also acceptable to spray the case directly, but only if you can do so without getting moisture on any of the cables or inside the louvers.

WARNING

Never spray anything inside the louvers. Doing so could short computer components and cause the computer to fail. When in doubt, spray the cleaning cloth, rather than the computer, to perform cleaning tasks. If nothing else, leave the dirty area intact, rather than risk damage to the computer.

Damp-Cleaning the Visual Elements

The visual elements of your computer include the monitor and any status displays, such as those commonly found on printers. These elements require a special spray made for computer monitors. You can obtain this spray from any store that sells HDTV supplies because the same spray is used on your HDTV. If you use regular window cleaner, you could potentially damage the plastic used on the front of the visual display, so make sure you use an appropriate spray.

Most people also use an exceptionally soft cloth to wipe the visual elements. A cloth of the right kind often comes with an HTDV cleaning kit. The extra you pay for such a kit will help ensure that your monitor lasts longer.

WARNING

Never spray anything directly onto a monitor or other visual element surface. Doing so will cause the surface to retain dirt in a way that makes it less usable. The dirt becomes embedded so that you can't see the underlying data as easily. This problem is especially noticeable with monitors. Make sure you use an appropriate cleaning cloth and spray the cleaning cloth, rather than the visual surface. The damp-cleaning cloth will help you remove dirt effectively without damaging the visual surface.

Checking for Damage

After you clean the outside of the system, it's important to check for damage. Damage can take a number of forms, but it's essential to look both carefully and thoroughly. Here are some examples of damage you might find:

- Frayed or damaged cables
- Broken visual elements

- Button damage
- Scuff marks (potentially from someone hitting the device with a push vacuum)
- Burn marks (potentially from arcing cables)

> **TIP**
>
> It may sound strange, but odors can tell you a lot about your computer. Burning components have a distinct odor. In some cases, you can actually smell a fan failure before it actually occurs. A damaged cable may also have a peculiar odor to it. So, breathe deep and see if you can smell something you can't quite see.

Cleaning the Inside

It's not required that you clean the inside of the case as often as you clean the outside. However, you do need to clean the inside from time to time. Dust can accumulate inside the case and your case may have filters that require cleaning. The dust that accumulates in the case can contain all sorts of things, including metallic elements. If you don't clean the dust out, some components could short. In addition, dust can cause your fans and storage devices to fail prematurely. The following sections discuss the techniques you use to clean the inside of your computer.

Opening the Case

Depending on how your case is made, you may have doors or simply a clamshell to remove. When a case has doors, make sure you remove both doors. In addition, remove the front bezel in order to access the fans, the front of the storage devices, and any filters contained in the front of the system.

Some cases will include fans on the doors like those shown in Figure 16-1. When removing the doors, you need to disconnect these fans from the system. Simply disconnect the power cable. Don't remove the fans from their holders on the doors. You want to keep the fans attached to the doors so that they don't get damaged.

Use a can of compressed air (such as the one shown in Figure 16-2) to clean the louvers completely. Spray from the inside of the case covering to the outside to make it easier to remove the dirt. You can use a damp cloth to remove any stubborn dirt as long as doing so won't potentially expose electrical components to harm.

Chapter 16: Maintaining the Hardware

FIGURE 16-1 Disconnect any fans that are attached to the door, rather than remove them.

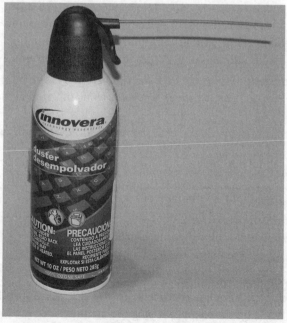

FIGURE 16-2 Use compressed air to remove dust and dirt from the case louvers.

> **NOTE**
>
> Make sure you use compressed air designed for computer or electronics use. This type of compressed air is dry and relies on inert gases that are less likely to cause damage to electronic components. In addition, the pressure is such that it won't damage components through excessive force.

> **WARNING**
>
> Discharge any static electricity from your body by touching the power supply case. The power supply case is grounded and will discharge even the small amounts of static that you can't feel, but could damage components. After you discharge any static, unplug the computer to avoid getting shocked. However, always treat components as if they're charged. Try not to move around while working—walking and other activities can build a static charge on your body that can damage components. When in doubt, plug the power supply in again and discharge any static before beginning work again.

Spraying the Dust Out

Use a can of compressed air (commonly referred to as a duster) to spray dust out of the case. Always spray from the inside of the case toward the outside whenever possible. Take your time and work carefully to avoid damaging any of the components.

> **WARNING**
>
> The outside of the compressed air duster will become quite cold. The best strategy is to spray in short bursts. If the can becomes damp from condensation, make sure you wipe it off to avoid getting moisture in the case.

It's easy to miss dust bunnies inside the case. Here are some suggestions for finding them all:

- Carefully move cables around and dust behind the cables.
- Dust out any empty bays.
- Blow out the power supply louvers.
- Use the plastic extension to blow behind objects.

> **WARNING**
>
> Do not ever use any kind of liquid inside your case. If there is some bit of dirt that won't come off simply by spraying the area with compressed air, leave the dirt, rather than damage the computer.

Chapter 16: Maintaining the Hardware

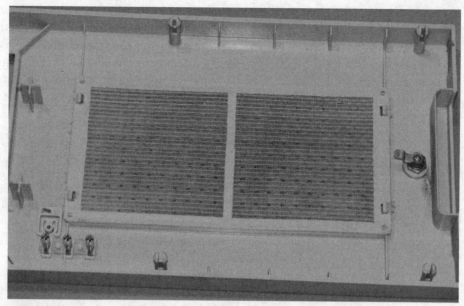

FIGURE 16-3 Look for all the filters that come with your case and clean them all.

Cleaning the Filters

Some computer cases come with filters. Remove the filters from the case (if possible) and blow air through them in the direction opposite of the normal air flow. Blowing in the same direction as the normal air flow will only serve to embed any dirt, rather than remove it. In some cases, the filter is actually part of an external case component, as shown in Figure 16-3. You should use the same procedure to clean this sort of filter.

> **WARNING**
>
> *Never wash the filters in your computer case. Moisture can remain after the washing process and cause damage to the system. Always use compressed air to clean the filters. If possible, try replacing excessively dirty filters with new ones. Online sources such as Amazon.com (see http://www.amazon.com/s/?keywords=computer+case+filter) stock replacement filters that will likely fit your case.*

Inspecting Your Hardware

It may surprise you to know that many people never actually look at their computer. After all, it probably sits under their desk and simply collects dust most of the time. Most people don't have any idea of what their computer even looks like because they may not have ever seen the entire box at one time. At most, they see the power switch long enough to turn the system on as needed. However, your custom computer represents a major investment, so you really do need to look at it and know the state of the components inside. The following sections provide you with goals for inspecting your computer and keeping it in great shape.

Checking the Add-on Boards

Inspecting your add-on boards will ensure that you don't end up with intermittent or unexplained errors. Nonrepetitive errors can be a real nuisance. They can masquerade as virus attacks or buggy software. They can also cause data loss and other problems. Here are some checks you should make:

- Ensure the add-on board is properly screwed to the case.
- Verify that the board is properly seated. Lightly push down on the board and it shouldn't move in the slot. (Don't actually wiggle the board because you can cause slot damage.)
- Check any cable connections to ensure they're secure.
- Move any cables that could interfere with fan operation.
- Perform any vendor-recommended inspection or cleaning tasks (most vendors don't provide any, but it pays to check the manual to make sure you aren't missing something important).

Keep in mind that you need to perform tasks carefully to ensure you don't damage the system. You must also observe proper precautions to prevent static discharge to the boards or electrical shock to yourself. The intent of the check is to look for potential problems before they occur.

Checking the Storage Devices

Storage devices come in two types: permanent and removable media. A hard drive is an example of permanent media—you can't change the media out. A DVD drive is an example of removable media—you can take the disk out and replace it with a new one.

In both cases, you want to ensure that any cables are properly attached. In addition, you want to ensure that any dust is removed from around the device and in crevices in the back of the device where the cables attach.

When checking a device that supports removable media, you want to take any media out and ensure that the mechanism works properly. It's also important to service the device with an appropriate cleaning product.

Chapter 16: Maintaining the Hardware

FIGURE 16-4 Use a special disk like the one shown here to clean your DVD.

For example, when working with a DVD, you can obtain a special disk to clean the lens as shown in Figure 16-4. The disk contains little brushes that remove dirt from the lens so the DVD can contain reading disks accurately. Likewise, tape players come with special cleaning tapes. The important thing is to remember that removable media devices are open and likely to get dirty inside, so you need to clean them.

NOTE
Most cleaning devices offer a limited number of uses. Track the number of times you have used the cleaning device and discard it when you've reached the maximum number of uses, even if the device looks like it'll continue to work. In many cases, it's not a matter of whether the cleaning device will work, but whether it's still clean enough to remove any dirt from the removable media device.

Checking the Cables

Cables can become twisted and damaged as the computer ages. A damaged cable may not fail to work completely. Sometimes a damaged cable experiences an increasing number of intermittent errors that are hard to locate. Verifying that the cable is firmly attached at both ends and in good shape can help you locate potentially bad cables.

It's also important to ensure the cables are out of the way so they don't impede airflow. You can use cable ties to tie up cables and put them out of the way, rather than have them dangling within the case. At the least, make sure you route cables in such a manner that airflow isn't impeded in any way and the case presents a clean appearance.

Keeping Things Cool

Clean louvers and fans that operate correctly help keep your system cool. A cool system lasts longer and works better. As your system ages, you can continue to clean the louvers, but the fans will eventually fail no matter how well you maintain them. The following sections describe two ways to ensure your fans continue to work as anticipated.

Spinning the Fan Blades

Use compressed air to verify that each fan can turn freely. Simply spray against the fan blades and the fan should move. Spray in such a way that the fan turns in the direction it normally turns to reduce the potential of fan damage. Make sure that the area around the fan is completely clean and clear of obstructions.

> **NOTE**
>
> *Some specialized fans may not turn freely, but still be perfectly usable. These fans use a motor that relies on digital pulses and provides precise output. The majority of these fans are mounted on the motherboard. A case, hard drive, CPU, or power supply fan will always move freely when sprayed.*

If you find a fan that doesn't rotate easily, replace it with the correct counterpart. A fan has specific characteristics that you must observe when obtaining a replacement:

- Size: Fans come in specific sizes. Most case and hard drive fans are 80 mm in size, but other sizes do occur. The fan size must match exactly or the new fan won't fit within the existing enclosure.

- **Voltage:** The voltage a fan expects is stamped right on the fan somewhere. The voltage must match exactly or you risk damaging the new fan and potentially the power supply or motherboard as well.
- **Speed:** Make sure you get a replacement fan with the same speed characteristics so that the part remains cool. If you can't find a replacement with exactly the same speed characteristics, use one that runs faster.
- **Static pressure:** Some fans are marked with a specific static pressure. These fans normally cool a device with fins, such as the CPU. The static pressure of the new fan you install must equal or exceed the static pressure of the old fan.
- **Cubic feet per minute (CFM):** Some fans are marked with a specific CFM rating. These fans normally provide case cooling. Replace the existing fan with a fan that has the same or higher CFM.

> **NOTE**
>
> *When using a fan that exceeds the current fan's characteristics, it's usual to expect additional noise. The new fan must provide a more aggressive blade configuration to move more air or a larger motor to run faster.*

Performing a Powered Test

After you have cleaned and checked the fans to ensure they spin freely, plug the computer back into the outlet. Start the computer just long enough to verify that the fans rotate as required and don't produce excessive noise (usually a sign of a failing bearing). Don't keep the computer on long enough to start the operating system boot cycle. Turn it back off as soon as you've verified that the fans spin as anticipated. Make sure you plug in any case fans that are attached to a door before you apply power to the system so that you can test them as well.

> **WARNING**
>
> *Never try to plug a fan in when the system is already running. The act of plugging the fan in can cause a surge that could cause system damage. If you forget to plug a fan in, make sure you turn the system off and disconnect the power first. Using this approach ensures that you can see whether the fan is working without damaging the system.*

If you find that a fan doesn't run when power is applied or the fan makes excessive noise, you need to replace it. The "Spinning the Fan Blades" section of this chapter provides tips on replacing a fan in your system.

Obtaining Spare Parts

Computer equipment fails at some point. Of course, your computer as a whole will last quite a while, so it pays to repair the broken parts. The following sections discuss how to work with spare parts to ensure you keep your system at peak performance at the lowest cost possible.

Understanding How Parts Fail

A statistical value, mean time between failures (MTBF), tells you how long the part should work on average. However, life rarely presents the statistical scenario and there is actually a mathematical basis behind it. Some amount of parts will fail early in their lifespan due to a wide range of issues that really are outside the scope of this book. After these early parts fail, the part as a class has a relatively long period where few parts fail. Then comes another cycle where the remaining parts die off. In other words, you end up with an inverted bell curve (called a bathtub curve) with most of the parts dying right at the point the MTBF says they will.

Fortunately, you can find out more about the whole issue of MTBF online. For example, you can find out more about the bathtub curve at http://www.weibull.com/hotwire/issue21/hottopics21.htm and http://www.weibull.com/hotwire/issue22/hottopics22.htm. This particular write-up is good because it answers a wealth of questions, such as what burn-in time is supposed to do. Then there is the whole discussion of whether it's better to buy a high-quality part and you can read about it at http://www.thesimpledollar.com/the-reliability-bell-curve-what-does-more-reliability-actually-mean/.

Creating a Parts List

Knowing what your computer contains at any given time is important because you don't want to have to open the case constantly to find out the specifics of your system. Having a parts list that contains the part number, serial number, part name, and vendor of each part in your system will save you considerable time and effort. Having a parts list that contains just these four items will let you order new parts for your system and determine when parts are outdated.

Some people include more information on their lists to improve the list's usability. These items can include:

- Installation date
- Software revision
- Firmware revision (a number that reflects the software version that is installed directly in the chips)
- Package list and location (the place you put the manuals and installation disks)

Shopping for Older Parts

If you really like your system as it is and some part fails early in the computer's life, getting a new part with exactly the same characteristics is quite doable. In fact, you'll probably get the new part for less than you paid for the original because the original is now outdated (even if it's as little as 6 months old). In most cases, you can simply go to the original store and get the part you need. Sometimes you need to go to a store that tends to sell older parts at a highly reduced price, such as TigerDirect (http://www.tigerdirect.com/). Some third-party vendors on Amazon.com (http://www.amazon.com/) will also sell older parts for a reduced cost.

There are situations where a part is unique and you can't obtain a newer version. For example, you might use a specialized sensor for some of your work and require an exact duplicate for future work. If none of the usual places sell the part you need, you can sometimes find the part at a liquidator store, such as A-Z Computer Liquidators (http://www.spintradeexchange.com/liquidation.htm). However, now you need to consider whether the part is used, new, or refurbished. Take time to figure out precisely what sort of part you're getting.

> **NOTE**
>
> *When shopping for older parts, it pays to comparison shop and to check on the store you're dealing with. Make sure you're not paying a high price for an older part that you could easily replace with something newer and better for less money. Older parts shopping brings the full meaning of caveat emptor (let the buyer beware) into play.*

Getting Updated Parts

When your display adapter or hard drive fails, it's a really good time to update your computer. The price of a new, updated part will likely be about what you spent on the original part and now your system will have additional capabilities. However, getting an updated part isn't always trouble free. Consider these questions when getting a new part:

- Can you actually install the part in your system? An updated part might require interfaces (such as SATA 3) that your existing system doesn't have.
- Will the driver work with your existing operating system?
- Are there known conflicts between your system and the new part? (You can get this information by reviewing forums and FAQs online.)
- Can your power supply handle the new device?
- Are there enough cooling fans to handle the new device?

Maintaining Essential Spares

Some parts fail often enough, are critical enough, and are generic enough that you want to keep some spares on hand, rather than having to order the part after it fails. For example, it's a great idea to keep extra fans on hand. Even though fans are pretty reliable, you want to have a replacement quickly when a fan fails. The reason is simple. If your system is running without that fan in place, a more valuable component could fail. Fans are cheap, generic, and essential to proper system operation, so it's not too much of a burden to keep spares on hand.

Another potential spare to maintain are cables. For example, SATA cables are pretty sturdy, but one could fail or you could accidentally damage one. In this case, the system won't run without the cable and you don't want to keep it offline waiting for the new part to arrive. Keeping a spare also makes sense in this case because you may decide to add another hard drive. The new hard drive will require a SATA cable to connect to the system.

> **TIP**
>
> *You can make spares out of parts that you update. For example, if you replace your perfectly usable display adapter with a newer one to obtain additional speed or functionality, you can turn the old display adapter into a spare. Yes, the spare won't work nearly as well as the new card, but if the new card fails, you can put your spare in place until a replacement arrives. You'll lose less time because the spare will keep your system running while you wait for the new part.*

It's important to keep the spares in their original packaging. Doing so will help maintain the viability of the spare. In addition, you can't always rely on a spare to function either, so you keep a spare for the spare. For example, if you feel a need to have a spare 80-mm fan, make sure you actually have two of them. That way, if the spare you have on hand is also bad, you'll have a backup. It's highly unlikely that both spares will be bad, so the two spares approach is usually all you need.

Creating a Maintenance Schedule

It's essential to maintain your computer on a regular basis. Mark the day on the calendar, write a note on your mirror, or do anything else needed to actually get the job done, but don't avoid doing it (as most people do until their computer completely dies from maltreatment). The frequency of maintenance tasks depends on a number of factors. Here are some things to consider as you create your schedule:

- How long you run your PC every day (more hours of operational usage means more maintenance time)
- The number of fans drawing air into the system (more fans mean more air being drawn into the case, which means more dust)
- Environmental factors (a dusty environment means more dust being drawn into the case)
- Heat generated by components (higher heat generation means more component stress and more maintenance)
- Number of component fans (the irreplaceable component fans need to be kept clean or the component will overheat)

Most systems used on a nearly daily basis will require quarterly cleaning as a minimum. A computer that is receiving heavy business usage may require cleaning as often as once a month. The outside of the case can be cleaned as often as once a week, depending on just how dirty the environment is and the number of fans in use. Observation is your best friend. If the computer looks dirty, make sure you clean it. Try cleaning the inside of the case once for each three or four times you clean the outside of the case.

17
Managing the Software

Software is the set of instructions that tell your computer how to perform specific tasks. When the instructions are old or simply incorrect, the computer can't perform the tasks you expect of it. This seemingly simple concept is the basis for a whole range of problems that drive users crazy. Because you've built your own computer, software issues will take on a new level of meaning for you. Managing your software means ensuring the software does what it's supposed to do all of the time.

The underlying purpose of software is to manage data. You might not even see the data all the time. For example, when you play a video game, it may not seem as if there is much data involved, the game tracks your settings and statistics for you. The point is, data and software are intertwined so that any discussion of software should also include a discussion of the data it manages for you.

This chapter helps you avoid a significant number of the problems that users can encounter by managing your software correctly. It's important to note that managing software can't eliminate every problem, but it can help you reduce the effects of even the issues that you can't control directly.

Performing Required Updates

Some people never perform any kind of update on their software under the guise of "If it isn't broke, why fix it?" The problem is that your software actually is broken—all of it. It's nearly impossible to produce perfect software. Someone, somewhere, will make a mistake and that mistake won't be caught—no matter how diligent the quality assurance (QA) department might be. This is the reason you need to perform required updates—to fix your broken software. Unfortunately, the fix isn't likely to make your software perfect either. Some bug will still lurk beneath the surface until someone finds it. The following sections tell how to perform software updates intelligently so that they actually do fix your software and make it better.

Considering What Broken Software Means to You

Broken software is a problem that affects everyone in a number of ways. The most common way that broken software affects you is in loss of functionality. You know that the software is supposed to work in a certain way, but it doesn't actually do so. Even worse is the situation where the software only fails on certain occasions, leaving you scratching your head as to precisely what went wrong. Errors that occur as the result of performing steps in a particular order are actually quite common, which is why the vendor will ask you for a set of steps to repeat the error as part of reporting an error.

There are more insidious problems with broken software. When software doesn't work as it should, outsiders can cause problems with it. For example, many viruses are based on the idea of the software not working as it should. A hacker will exploit the error and use it to do something nasty on your machine. An amazing number of software updates you receive are specifically for the purpose of thwarting hackers. Unfortunately, the hackers are quite adept at their trade and keep coming up with new ways of getting into your machine through software errors.

It's also important to understand that broken software can cause problems with your data. An error can damage your data in ways that make it hard to recover. Of all the parts of your system, your data is worth the most because it's inherently hard to recover lost data or to fix broken data.

Setting Software to Check Updates Automatically

Most software today provides some sort of automatic update setting. In some cases, you really don't have a choice in the matter; the software is going to check for updates whether you want it to do so or not. Some software actually performs the updates without your permission, which can be a problem for a number of reasons:

- The update may not actually apply to you and adding it to your system causes problems.
- An error in the update actually makes the software worse, not better.
- Performing the update at a time you aren't prepared for it can result in unforeseen problems, including data loss.

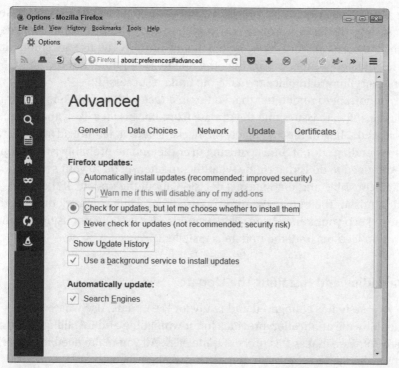

FIGURE 17-1 Make sure you check updates automatically, but apply them manually.

The best possible setup is one in which the software notifies you of the update and then lets you decide when to apply it. Of course, this means you have to take responsibility for actually applying the update in a timely manner. Because some users aren't performing the updates ever, some software vendors have chosen to make forced updates automatically.

In most cases, the choice of how to perform updates appears somewhere in the application options or setup dialog. For example, Figure 17-1 shows the Options dialog for Firefox (a browser that runs on most platforms). In this case, the updates are configured such that the application does check for updates automatically, but lets the user determine when to apply the updates.

Notice that Firefox provides other update settings. For example, you can choose to use a background service to perform the update. Relying on a background service makes the update less noticeable—you can continue to work while the update takes place. In this case, the application also performs updates of the search engines automatically.

Validating the Update

When you receive notification of an update, make sure you take time to research it. If the update is broken, you can be sure someone will be talking about it. It's also a good idea to determine precisely what the update will do for you. In some cases, such as the Firefox browser, the update could actually disable add-ons that you need to perform work. Make sure that any add-ons you use are compatible with the update. In short, performing the update is required at some point, but be sure everything is in place to make the update successful.

It's important to remember that you're in a race with the hackers who want to exploit the software issues that the update fixes. Because it's likely that a hacker discovered the bug in the first place, the hackers have a definite advantage over you, so spending a lot of time agonizing over the update probably isn't an option. What you need to do is determine what the update will do to your setup as quickly as possible and then make a decision about the update. If there are workarounds for the bug, make sure you use them while you work through the update to keep your system safe. Workarounds may include not using certain application features until the update is installed.

Downloading and Installing the Update

Once your system is configured and ready for the update, use whatever functionality the application provides for downloading and installing it. Most modern software makes this process quite easy. All you really need to do is click a button when asked about the update to start the process.

> **TIP**
>
> *Always shut down any running applications when you perform an update so that as many system resources are freed as possible and the update has the best possible chance of completing without encountering some type of conflict. The express setup option for an update is usually the best choice unless you have advanced application knowledge and really do need to perform a custom setup or the update fails when using the express option. Keep things as simple and uncluttered as possible when performing an update to ensure the update succeeds and your application actually runs afterward.*

Some older and less complex software may require that you perform a manual update. In most cases, the vendor will provide you with a series of steps to follow to ensure the update is successful. Make sure you read and understand the full set of steps before you perform any of them. If you don't understand something, ask questions before you start the update. Otherwise, you could find yourself in the middle of the update process with a machine that won't respond to commands. (For this reason, you also want to print out the update instructions to ensure you have a copy available for your use.)

> **WARNING**
>
> Never download updates from a site other than the one specified by the vendor. A lot of third-party sites will likely have the update file as well, but you can't be sure these sites haven't contaminated the file in some way. A lot of hackers will piggyback virus or other code onto an update that you download from a third-party site.

Testing the Update

A huge mistake that many people make after applying an update is assuming that the update actually succeeded. After all, the dialog box at the end of the update process told you that the update was successful. Unfortunately, what that dialog box is actually telling you is that the update software ran without detectable error—not that the update actually succeeded. In order to verify that the update was successful, you need to test the application afterward.

Always start with a copy of your data, rather than the original data files. Updates can cause all kinds of problems, including data damage. Here are some steps you can use to test your update:

1. Start the application and perform some tasks that you normally perform using the software.
2. Load the test data and play with it, rather than your original data. Try all sorts of tasks, just to make sure the update is stable.
3. After you play with the update for a while, save the data file and shut down the application.
4. Restart the application and load the test data again. Verify that your changes took effect as expected and that the application is running properly.

Keeping Things Secure

Unless you want everyone in the world to know your personal information or your system to run as slow as molasses in January, you need to keep it secure. Security is a complex topic that can require entire books to cover in any depth. Yet, there are some basic security measures you can take to keep your custom system running well. The following sections discuss some of the measures you can take to secure your system without a lot of hassle.

> **WARNING**
>
> Some people think that once they secure their system, nothing can get through. Security measures are like locks on your house—they keep

honest people honest and really poor crooks out. If someone really wants to get into your house or your computer, they can do so. Of course, you'd instantly notice that someone has rifled through your things if a crook breaks into your house. Likewise, watching your computer closely can help you detect the presence of a hacker on your system. Vigilance is a required part of securing your system.

Starting with Passwords

Password protecting your system, the software you use, and the sites you visit is important. You can find a wealth of advice on just how to do it too. Most of the advice is meant to be helpful, but sometimes it ends up being confusing instead. Some advice is simply too hard to follow. Like the exercise routine you decided to try and eventually dropped because it expected much and delivered little; complex security schemes usually fail. Here are some things you can try to make working with passwords easier:

- Choose a different password for each site and then store that password in a password vault or use a password you can associate with the site to make it easier to remember. (The *PC Magazine* review at http://www.pcmag.com/article2/0,2817,2407168,00.asp provides you with a list of some of the better password vaults available.)
- Use a passphrase because it's going to be just as hard to guess as a complex password. A password such as $%A1bz22F is really hard to remember, so it's probably not the best choice. Use a passphrase such as My D0g Sm1les Pretty! (notice that the o is replaced with 0 and the i is replaced with 1). Using single word passwords is going to cause you woe because hackers have too many techniques available to guess them.
- Keep your passwords and login name private. No one else needs to know what they are. Even an administrator has means other than using your login name and password to access applications, sites, or other software you need to work. Requests for your login name and password are usually made by hackers who want to gain illegal access to your system.
- Choose passphrases that don't identify you or anything about you. For example, a passphrase of "My B1rthday 1s May 1st." isn't a good passphrase because it contains personally identifiable information that the hacker could guess. Using "The Cl0uds Ar3 Blu3?" instead is better because a hacker can't easily guess it based on any knowledge gained about you.

Installing the Correct Background Applications

Securing your system, the act of applying locks in the form of security and keys in the form of passphrases, is a good step. However, the best house-security systems include breaking glass sensors, motion detectors, and switches to detect both open doors and windows. Likewise, your computer needs the equivalent of detectors to ensure someone doesn't force their way into the system. With this in mind, you need to install these kinds of software onto your system:

- Firewall: Detects unwanted incoming and outgoing data.
- Antivirus: Determines if your system is under virus attack and thwarts it.
- Application monitor: Keeps track of application activity and alerts you when something unexpected happens.
- Content checker: Verifies that any content you download or access online is data only and not an executable that will damage your system.

There are other sources of sensors you can install on your system, but these are the most common. Fortunately, you can often obtain everything needed with a single package, such as AVG. The nice thing about some packages like AVG is that they have a free version you can try to see if you like it (http://free.avg.com/us-en/homepage). You can find a review of free antivirus packages that include other sorts of detectors on the *PC Magazine* site at http://www.pcmag.com/article2/0,2817,2388652,00.asp.

Applying Security Updates Quickly

Standard application updates have a level of urgency because a hacker can use them to cause problems on your system. However, security updates should always receive the highest priority. An update that is marked as a security update points to a specific problem that hackers may already use to obtain access to your system. Some vendors, such as Microsoft, are careful to mark security updates as shown in Figure 17-2 so that you can give them higher priority. When working with other vendors, it's important to read about each update to determine just how it affects your system and whether it includes some form of security fix.

Some security updates affect zero-day vulnerabilities (or attacks). A zero-day vulnerability is one that hackers are already using to access systems, so it's not a matter of if a hacker will use the vulnerability, but when the hacker will attack your system. Giving zero-day vulnerabilities immediate priority is essential if you want to keep your system safe. You can read more about zero-day vulnerabilities on the Symantec site at http://www.symantec.com/threatreport/topic.jsp?id=vulnerability_trends&aid=zero_day_vulnerabilities.

Reading Those Installation and Update Screens

One of the biggest potential holes in your security setup may be the user's ability to disregard risks of all sorts. For example, users will often give out their login name and password to a perfect stranger over the phone when asked. However,

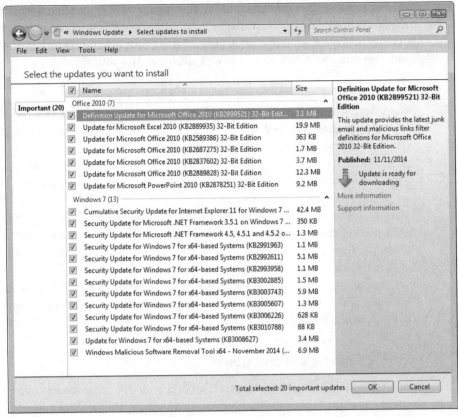

FIGURE 17-2 Always give security updates a higher priority.

even more common is clicking through setup screens without even looking at them. This particular failing even affects experienced users. It's easy to look at a setup screen used to install or update an application and not see the little checkbox at the bottom asking to install some third-party product that's likely packed with viruses, adware, and Trojans. When viewing a dialog box, make sure you read the entire dialog box to avoid potential problems with the installation.

The reason that you find these add-ons provided with certain categories of software is to pay for the software. There truly is no free lunch. Instead of you paying for the software, the vendor has a third-party to pay for it. The add-on is the cost to you of having the third-party vendor pay for development of the product. The primary vendor is betting that enough people won't read the screens to make creating the software profitable.

WARNING

Unfortunately, once you have one of these third-party applications installed, getting rid of them can be daunting. In many cases, you can

Chapter 17: Managing the Software

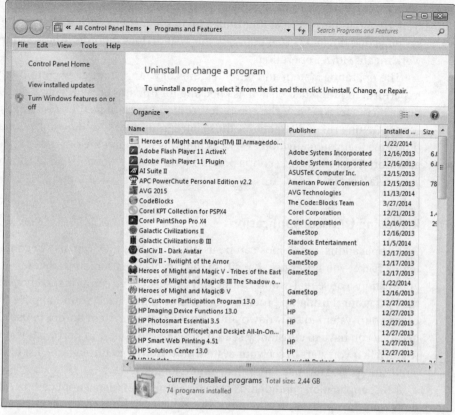

FIGURE 17-3 The Programs and Features console of the Control Panel provides a list of applications.

remove the software by locating the software's directory on your hard drive and using the uninstall program to remove it. When working with some software, you can sometimes find the third-party application using some nebulous name in the Programs and Features console of the Control Panel shown in Figure 17-3. However, some of the software will require manual removal techniques that you have to find online (usually not on the third-party vendor site).

Dealing with System Slowdowns

After some period of time, you'll notice that your system is slowing down. No, it doesn't have anything to do with dirt in the case or the aging of components. The source of the slowing is actually easy to understand. As you add applications, the hard drive becomes disorganized (causing the operating system to do more work to obtain files), and clutter builds up in the form of settings, hence the system slows down. Every bit of work the system does requires time, so more work means slower operating.

The operating system and other features of your system need computer time to perform required tasks, such as checking for updates or retrieving a file from disk. Of course, you want the computer to spend most of its time servicing your needs. Getting rid of distractions, those background tasks and organizational problems that you really don't need, is one way to obtain the operating system's attention. The following sections provide some basic guidelines on making your system fast again.

Removing Unused Applications

Free applications and demos can be useful. However, they can also become that tennis racket you never use that sits in the closet. The problem is that these applications cause real-world clutter and can also have components that sit in the background using up processor time. Clutter and background applications cause your system to slow down and also make your system harder to use because you have to constantly search through the clutter to find what you really want. When you start to see your system slow down, go through the list of your applications and remove those that you no longer use.

Most applications and demos provide a simple uninstall application. All you do is start the application to remove the application from your system. Some operating systems, such as Windows, also provide a centralized listing of applications. On Windows, you can find this list in the Programs and Features console of the Control Panel. Simply search through the list to find the applications you want to remove.

> **TIP**
>
> *Many uninstall programs will give you a choice of removing the application directory or saving the data files and settings for the application. If you plan to use the application at a later date, saving the data files and settings can save you time and effort. However, when you don't plan to use the application again (or even if you're in doubt), removing the application directory is the best idea.*

Cleaning the Hard Drive

When you initially buy a system, the hard drive is completely clean. There is no data located on the hard drive. As a consequence, the hard drive provides the best possible access times because every file is stored on one contiguous location.

The hard drive doesn't spend any time looking for the pieces of a file—all it does is read the file in its entirety. However, as the hard drive continues to see use, the number of contiguous spaces gets smaller and smaller, forcing the operating system to store a piece of a file here and a piece of a file there. The loss of large contiguous spaces is called fragmentation. In order to regain the hard drive's speed, you need to defragment it. Here are some sites that discuss how to defragment drives for specific operating systems:

- Windows 7: http://windows.microsoft.com/en-us/windows/improve-performance-defragmenting-hard-disk
- Windows 8: http://anewdomain.net/2013/08/07/how-to-optimize-your-hard-disk-in-windows-8-and-defrag-burgess/
- Mac: https://discussions.apple.com/docs/DOC-4032
- Linux: http://www.linux-magazine.com/Online/Features/Tune-Your-Hard-Disk-with-hdparm

It's important to remember to actually clean your hard drive from time to time by removing data files and directories you no longer need. It may seem as if hard drives have infinite storage space today or that you can simply keep adding hard drives as needed, but hard drives that are cluttered with data that you no longer require can cause serious problems. If nothing else, archive the information to tape or a DVD for later retrieval.

Cleaning Up the Operating System

In some cases, you must also consider how to clean up the operating system. For example, Windows gets cluttered with services that you really don't need. It also has the infamous registry, which keeps growing as the operating system works with applications and updates. Of course, you can perform this work manually, but it's generally a bad idea to do so because one mistake can make your installation nonfunctional.

What you really need is a cleanup utility such as CCleaner (https://www.piriform.com/ccleaner) that makes the work easy and less prone to error. Fortunately, there is a version of CCleaner for both the PC and the Mac. The article at http://hubpages.com/hub/5-Free-Tools-that-will-clean-your-PC-like-a-PRO tells you more about tools you can use to clean Windows up and the article at http://thetechreviewer.com/tech-tips/best-mac-applications-cleaning-speeding-up-apple-computer/ provides the same information for Mac users. Linux users will have to rely on a tool that's a little more manual in nature. You can read about some of these tools in the TechRepublic article at http://www.techrepublic.com/blog/five-apps/the-five-best-linux-file-system-cleaning-tools/.

Keeping Data Safe

Even though your machine is expensive and your time valuable, your data has the greatest importance. Your data represents the wealth of many days' work and it can't easily be replaced. In fact, there are situations where you can't replace it at all. The worst case scenario is where you can't replicate the data either and must simply view it as being gone. Some organizations have gone bankrupt over the loss of data because data represents such a huge investment in time and effort. The following sections provide a quick overview of some of the tasks you need to perform to maintain the safety of your data.

Performing Backups

A backup stores a copy of your data on some other media, normally in some other location. The purpose of a backup is to provide a copy of your data should the original copy become lost, stolen, corrupted, or otherwise not usable. Of course, a backup is only worthwhile if it's current. From the moment you complete the backup, the data begins to age. Every moment it ages is another moment you must make up during a recovery, so frequent backups are essential if you want to recover from a disaster quickly.

There are many kinds of backup (tape, DVD, and cloud being the most common), each of which has advantages and disadvantages. This chapter can't cover them all or discuss them in detail. However, a DVD backup has the advantage of being fast (both backup and restore) and relatively small. In addition, a DVD backup is relatively stable and less likely to become damaged than other forms of backup, but it has a short shelf-life. A DVD backup is usually viable for around 5 years, after which the media begins to degrade and you start losing data. Another issue is that DVD backups are limited by the amount of space a DVD can store, which means you usually need multiple DVD disks to complete a single backup.

Tape backup has the advantage of being able to store large amounts of information virtually forever when the tape is stored correctly. The actual length of storage depends on the tape, but storage times of 30 years are typical. A longer storage time is impractical because the technology used to interact with the tape will become outdated long before then. The disadvantages of tape include a need to ensure the storage environment protects the tape (not too hot or too cold, and within the correct humidity range) and slow read times.

Cloud storage depends on sending your data across the Internet to an online storage facility. The speed at which this happens depends on the storage facility capabilities and the speed of your connection. The longevity of the backup varies. However, cloud storage has the advantage of being relatively inexpensive, capable of storing anything you have to backup, and automatic (depending on the software you use).

> **WARNING**
>
> *It has become popular to use cloud backup. Unfortunately, cloud backups come with all sorts of issues, most of them hidden from view by the vendors who support the cloud backup software. The blog post at http://blog.johnmuellerbooks.com/2012/02/07/saving-data-to-the-cloud/ describes some of the issues using cloud backups. The blog post at http://blog.johnmuellerbooks.com/2014/09/05/an-issue-with-cloud-computing/ expands on some of the problems you'll encounter. Yes, cloud backup is better than no backup at all, but be sure you perform a cloud backup knowing the risks of doing so.*

Keeping Backups Off-Site

Backups are only useful if they're undamaged. Fire and flooding tend to be indiscriminate. The same fire or flood (or other disaster) that destroys your computer could easily take your backup with it. So, the backup becomes a false sense of security that you can't really use because it's simply not available. To keep this from happening, you need to put the backup somewhere else to keep it out of harm's way.

Unless you're using cloud backup (where the backup is stored somewhere else by definition), you need to come up with a plan for storing your backup media somewhere other than your office or home. Many people rely on a friend's house, a safe deposit box at a bank, or a facility specially designed for the purpose. In fact, there is at least one person who stores their backup in a mailbox outside their home. The mailbox idea is better than nothing, but if you live in a climate like mine, it's probably not the best solution.

Encrypting Data as Needed

If your data is sensitive in any way or there is a legal requirement to do so, always encrypt it. The advantage of encrypting your data is that no one can read it without the key, which should be robust. The disadvantage is that encrypting your data causes the data to take up more space on the backup media. Most backup software offers you the choice of encrypting the data or compressing it. Compressed data requires considerably less media and backs up far faster than uncompressed data.

Overcoming Disasters

You're going to experience a disaster at some point. There will come a morning when you turn on your system and nothing happens. It's also possible that you'll be working along, smell a whiff of smoke, and suddenly find yourself staring at a blank screen. In fact, there are many ways in which your computer can suddenly become a boat anchor—pretty much not usable for anything. Fortunately, most of these issues aren't permanent. With a little patience, you can usually figure out what's going on, fix it, and get right back to work again. The following sections provide information on some of the most common sources of disaster and some suggestions for fixing them.

Checking Drivers, Services, and Agents

One of the most elusive problems you can encounter is an errant driver, agent, or service. Drivers are the worst because you absolutely have to have the driver in place to access the hardware. When a driver goes bad, it can look like the hardware is bad or it can look like a particular application is working incorrectly.

> **NOTE**
>
> *Driver, agent, and service problems sometimes make no sense whatsoever. In a real-world scenario, I actually encountered a system where the backup agent caused the camera to malfunction when you plugged the camera into the Universal Serial Bus (USB) port. The system would actually freeze if you tried to access the data on the camera. I fixed the problem by stopping one agent at a time until the camera suddenly started working. It turned out that the backup agent thought the camera was backup media and instantly grabbed hold of it for backup purposes.*

Working through this particular problem can be tricky. You need to be patient and try just one thing at a time. If you start trying multiple fixes at once, you'll never know what caused the problem and it will recur, usually at the worst possible moment. The following sequence will usually help you locate and fix driver, agent, and service problems.

1. Update the agent, driver, and service software one item at a time.
    - Start with agents (applications that run in the background and monitor the system) because they're the easiest to access and the least likely to trash your system.
    - Update the drivers next because most vendors make driver updates easy.
    - The services can be a problem because most vendors don't even let you know they exist, but try to update them as you can.
2. Stop the agents one at a time and see if you can detect any differences. Restart only the essential agents (a backup agent isn't essential, but your virus software and firewall should be restarted as soon as possible to keep your system safe).

3. Carefully stop the services one at a time. It's essential that you don't stop critical services (which can be horribly difficult to determine). If in doubt, leave the service running until last, research it online to determine its criticality, and turn it off only if it's possible to do so. Some operating systems make it difficult or impossible to stop critical services—don't try to override this protection.
4. Uninstall drivers one at a time. Removing the driver will make the associated hardware inaccessible, so this step isn't possible for critical hardware like the display adapter, motherboard components, and keyboard. Reinstall the driver after you test it.

Testing Errant Hardware

Throughout this book, you've assembled hardware and then tested it afterward. So, it shouldn't surprise you much to find that you already have many of the skills required to check your hardware for problems. However, now you're dealing with a completed system that has software installed and may not quite react the way that it did while you were building it. Here are some things to do when dealing with potential hardware errors:

- Boot the system and carefully read the information that the boot process provides. You can often see the information better by pressing the key (normally Delete) that accesses the machine setup during the boot process. The system will typically display messages or even tell you (using spoken words) what it thinks the problem is.
- Check for the obvious, such as cards that aren't seated properly or cables that aren't making good connections.
- Don't replace parts indiscriminately. Think about the piece of hardware that's most likely to fail. For example, display adapters and hard drives have moving parts. Yes, that fan on the display adapter really does count as a moving part and it's likely to fail faster than something that isn't moving. Check devices with moving parts first and always focus on the device most likely to cause the failure.
- Look for obvious signs of damage in your system. It doesn't happen often, but sometimes you can actually see what has gone wrong.
- Switch the part you think has failed with a spare part or a similar part from another machine to see if the new part fixes the system.

Restoring a Backup

In some cases, you can't actually find the issue that is causing problems, but the hardware checks out and the system does finish the hardware boot process as a minimum. Perhaps the operating system boots as well, but you might not be able to get beyond that point or the system may act flaky after it does come up. There are situations where some damage has occurred to the data on your system and that damage is causing the problem. To fix the problem, you may have to restore the backup copy of the data and reboot. However, given that the backup is out-of-date and you'll lose information using this approach, you need to use it carefully.

> **TIP**
>
> *Windows has a restore point technology that can help you uninstall some changes made by the operating system or update software. There are also ways to restore individual files. Look for operating system features that can help you recover without restoring everything before you rely on the restore process to fix the problem.*

There are times when you must take drastic action to fix your system. This is the last ditch, nothing else worked, solution to the problem that some people try first. Make sure you try everything else you can think of before you try the following set of steps because most people find that they're dissatisfied with the results because the system isn't quite the same as it was before when you finish.

1. Make absolutely certain that you have a good backup (assuming the system is in a condition to allow you to perform this check—often it isn't).
2. Reformat the hard drive.
3. Reinstall your operating system, applications, and any required updates.
4. Restore your data from the backup.
5. Test your system for any deficiencies and correct them.

18
Preparing for Updates

P arts age at different rates, some have extremely long lifespans. When you build your own PC, you have the chance to reuse these parts in updated systems. Other parts don't have a long life or you find that they require updates long before they're toast due to other issues. For example, one of my systems still has an 80486 processor in it. I use it to track my movies and play really old games, but its useful life was over a long time ago. The fact that the system even boots is amazing. On the other hand, one of my systems still has a serial port card in it from a long time ago. As long as the serial port continues to fit in a system that I can use, I'll continue using it as an extra port. The point is, some parts do last a long time.

When you buy a PC at a store, you normally throw out the whole system—the good with the bad, so you can't optimize your purchasing power. Building your own PC has the advantage of reusing components. For example, if your case is in good shape and you like your case, then there is no reason to replace it. Optical drives have long lives and as long as your motherboard supports the interface, there is no reason to replace it. This chapter is all about updating your existing PC so that you can get the equivalent of a really cool new machine at a significantly reduced price.

Issues to Consider When Reusing Off-the-Shelf PCs

It would be easy to think that an off-the-shelf PC must simply be a PC that someone built in a factory setting using the cheapest available parts. That's only partly the case. Upgrading an off-the-shelf PC is always a risky business and will likely cause problems for you. It's so much hassle in some cases that many people really do just throw out the old system, rather than try to recycle any of it. In general, you find that these components are upgradeable and reusable:

- Display adapter
- RAM
- Power supply
- Storage devices
- Fans

In order to save money, the vendor who builds an off-the-shelf PC looks for the lowest price parts possible. The reusable parts could be generic versions of name brand parts or they might simply be cheap parts. For example, your display adapter may use a name brand chip, but come on a board that uses less expensive companion parts. It pays to look at any parts you want to reuse with extreme care before you rely on them in a new system.

Some parts are likely to use a custom setup and may not even fit in a standard case. In fact, in a few cases, I've even found that the case is custom to the point where a standard motherboard won't fit. You shouldn't rely on these parts fitting in a new system or being upgradeable in most cases:

- Motherboard
- Add-on cards
- Case
- Specialized external port setups

Maintaining a Wish List

When you built your PC while going through this book, you likely needed to cut a few corners in order to meet your budget. The items that you left out of your system aren't gone for good; you can add them to a wish list for later inclusion. In fact, it's helpful to maintain a wish list for your system so that you can keep focused on the upgrades that you want most when you see the sales brochures arrive in your e-mail. Buyer's remorse is a common problem for people who build their own systems and become infatuated with a sale item that really isn't an essential item for a system.

Create a wish list for your system based on priorities and general part type. Include the parameters for the upgraded part. For example, if you want to add another 16 GB of RAM to your system, your wish list should include the RAM specifics, such as bus speed. Always buy matched RAM for your system. In fact, it's actually a good idea to buy RAM from the same vendor when you can find it.

Some parts are generic in nature. For example, when you want to upgrade your power supply so that you can support an additional display adapter, the main concerns are the power output of the power supply, whether the cabling is shielded, and the type of fan the power supply uses. Getting a power supply from a top vendor, such as Firepower Technology, is always a good idea, but not absolutely essential. Power supplies use the same screw hole setups, provide the same sorts of connection cables (although, higher quality and higher wattage power supplies tend to provide more connection cables), and have approximately the same size box, so it's easy to substitute one type for another.

A few of the parts on your wish list are going to be quite specific. For example, if you currently have a single Scalable Link Interface (SLI) or CrossFire compatible display adapter and want to add a second one, you must get precisely the same model from the same vendor to ensure success. The problem for most do-it-yourself builders is that specific components can go out-of-date quite quickly, so they should appear at the top of your list even if they're not the highest priority item from the perspective of need. It's better to get the part a little too soon, rather than not get it at all.

> **TIP**
>
> *Second and third components are often best bought just at the point where the vendor is starting to dump inventory to make space for an updated component. You can get the component at a greatly reduced cost and still be certain it will match your existing component. In addition, the component you get will likely come with the latest firmware. The drivers and associated software will be mature at this point and present you with a solid, reliable solution for your system that will last for quite a while.*

Building with Expansion in Mind

When creating your plan for your PC, you obtained parts, particularly the motherboard, with expansion potential. Expansions add to the functionality of your system and let you do more with it. Normally, expansions rely on components that were too expensive or possibly unavailable when putting your system together. Every expansion you perform requires careful planning because the state of the PC industry changes so quickly. A component that was new and viable last year may be unusable during an expansion the following year. The following sections provide advice on performing the most common system expansions.

Developing a Processor Plan

The documentation for your motherboard includes a list of processors that the motherboard can accommodate. If you're like most PC builders, you get a motherboard that includes processors in the list that you can't afford during the initial building process. Adding a new processor is an easy way to obtain improved speed from your system. You update the processor by removing the old processor and then installing a new processor using the technique described in Chapter 5.

> **WARNING**
>
> *Make sure you follow proper procedures when working with electronic components, including discharging any static electricity that may have built up on your body. Even if you can't feel the static discharge it can still damage components, causing them to fail when you try to power them up. As components have become smaller, they've also become more sensitive to static discharge. The easiest way to discharge static electricity is to plug the power supply in, discharge any static by touching it, and then unplug the power supply so you can work with the components safely.*

There is a twist to upgrading processors that you need to think about. Your motherboard has a specific socket type. Any processor that supports that socket should fit on your motherboard. Between the time that you built your original machine and the time of the expansion, the processor vendor may come out with a new processor that will fit your motherboard. If this is the case and the new processor provides features or speed that you really want, then use this process to ensure the new processor will actually work.

1. Check the motherboard vendor site for updates. New processors often require an updated basic input/output system (BIOS). Make absolutely certain that you follow the vendor upgrade instructions precisely because you generally get just one chance to do it right.
2. Verify that the new processor will actually work with the motherboard with the BIOS upgrade in place. The motherboard vendor should provide you with this information, but often the motherboard vendor hasn't tested the new processor. The processor vendor may be able to help as well. In some cases, you can get peer support, but relying on peer support can be quite tricky (you don't even know the person you're talking to has actually tried the update).
3. Before buying the new processor, check the return policy. You'll likely find that you can't return the processor under any circumstances, especially not if the processor has been placed in a motherboard and tested. In short, it pays to know that the processor will work before you buy it. If you're absolutely certain the processor will work, buy it from your favorite store.
4. Save the old processor when you remove it. Carefully package it in anti-static material or in the packing material used by the new processor. You may have to reinstall the old processor if the new one doesn't work.
5. Install the new processor using the technique described in Chapter 5. Test your system thoroughly to ensure the new processor actually works.

Upgrading Your RAM

You can't install too much RAM. At least, you'd have to try really hard to do it. A RAM upgrade will usually provide a speed boost by reducing the time the operating system spends swapping items in memory to disk. In addition, it can make your system more reliable by keeping items in a specific place in memory—reducing the risk of errant reads by applications. In short, RAM is an important addition when you can make it.

Before you can expand RAM, you need to power the system down and ensure you have additional memory slots. Most systems today come with two, three, or four memory slots, but motherboard configurations do vary. After you verify there is a memory slot to use, you can buy the RAM you need. It's absolutely essential that you obtain the correct type of RAM and that the bus speed matches the bus speed of your existing RAM. It's even better when you can get the RAM from the same vendor.

> **NOTE**
>
> *Make sure your operating system will actually support the RAM increase. When you start getting into higher amounts of RAM, an older operating system will refuse to see the RAM, even though the system says that it exists and is fully functional. Both the hardware and the software on your system must be able to support the RAM increase.*

Once you have your new RAM in hand, you can install it using the technique found in Chapter 5. Make absolutely certain that the new RAM is seated properly. Verify during the startup process that the system actually sees the new RAM. You may need to enter the CMOS setup to ensure that the hardware sees the RAM. After you boot your operating system, make sure the operating system sees the RAM increase as well. It's not automatically assumed that the operating system will see the RAM increase.

Upgrading a Display Adapter

Display adapters are expensive. In fact, they can easily outpace the cost of the processor in most systems. However, display adapters also lose value rather quickly and you can often get last year's display adapter at a considerable savings. When getting an updated display adapter, you need to ensure the new display adapter has the same bus type and your power supply has enough capacity to support the new display adapter's power requirements. In addition, you need to ensure that the new display adapter will work with your existing operating system.

> **NOTE**
>
> *The operating system may come with a generic driver that will work with your new display adapter, but this generic driver won't use any of the display adapter's advanced features. In addition, the display adapter will run slower than if you use a vendor-specific driver. Make sure the display adapter vendor supports your operating system of choice.*

Once you have the new display adapter in hand, you can install it on your system. In most cases, you must first remove the existing display adapter driver for the old display adapter and install the generic display adapter for your system. Otherwise, your operating system may fail to boot when you restart the machine. Install the display adapter using the technique described in Chapter 7 of this book. Part of this process will install the new display adapter driver specifically designed for your new display adapter and the operating system of your choice.

Adding a Display Adapter

If you have a SLI or CrossFire configuration, you can add a second or sometimes a third display adapter to your setup. The additional display adapter works in tandem with your existing display adapter to increase graphics processing speed and to provide additional graphics memory for display objects such as sprites. In addition, you gain additional ports so that you can connect more monitors to your system. In short, adding another display adapter increases graphics capability quite a bit.

Before you buy a second display adapter, check your power supply. Make sure the power supply is configured to support two display adapters of the type that you want to use. In most cases, new power supplies support both SLI and CrossFire configurations, but it pays to check first. You must also consider the power requirements for the new display adapter. The power supply may not provide enough power to support the second display adapter. Remember that you need to provide enough capacity to support surges—when the display adapter is called upon to do more than the usual amount of graphics process (such as when playing a game).

When adding a second display adapter, you normally don't need to worry about the operating system driver. The driver is already configured to use multiple display adapters when installed. In fact, when you boot your system after installing the display adapter, you probably won't have to do anything to start using the additional features the second display adapter provides.

> **NOTE**
>
> *When working with a motherboard that supports more than two display adapters, make absolutely certain that you put the second display*

Chapter 18: Preparing for Updates

adapter in the correct slot. Otherwise, the motherboard may fail to recognize the new display adapter.

However, the hardware installation can be time consuming and somewhat confusing at times. To make the two display adapters work together, you must install the second display adapter in the correct slot and configure the motherboard appropriately. Every motherboard manufacturer seems to have a different process for performing this task, so you must read the configuration technique in the motherboard manual. Make sure you read the entire configuration procedure and understand it fully before you begin the process of installing the second display adapter.

Increasing Your Storage

Of all the expansions you can perform, adding another hard drive, DVD, or other storage device is the easiest. In general, you don't have any special drivers to install—the motherboard and operating system recognize your new device automatically. Chapter 8 tells you about working with permanent storage devices.

You may wonder about potential problems when working with a new storage device. The only thing you really need to consider is whether you have another cable of the right type to connect your new storage device to the motherboard or a host adapter on an add-on board. The hard drive never ships with the required cables. You should also check your power supply capacity, but the power supply is usually not a problem because storage devices have relatively low requirements.

Obtaining Special Add-ons

There are a lot of special add-ons that this book simply doesn't have space to cover. It's possible to add just about anything to a PC today. However, special add-ons fall into categories that are covered in other places in the book. They usually don't require that you develop a new set of installation skills. In fact, it's in the vendor's best interest to make any specialty device as easy as possible to install. With this in mind, you normally install a special add-in as one of the following items:

- Universal Serial Bus (USB) port plug-in: This is the most common approach. For example, the test system supports both Wi-Fi and Bluetooth through a USB port plug-in.
- Video port plug-in: Some devices rely on a High-Definition Multimedia Interface (HDMI) or other video plug-in to interact with your system. To support such devices, you need a display adapter or other add-on board

that supports them. Motherboards don't typically include HDMI ports as a feature.
- Other port plug-in: Even though other types of ports have become less popular, some devices still use them. For example, it's possible to find devices that still require the RS-232 or RS-422 serial port interfaces. Older scientific devices often fall into this category.
- Add-on board: Some add-ons require an expansion slot on the motherboard. The problem is that motherboards have fewer expansion slots than they once did. Before you get an add-on that requires an expansion slot, make sure you have a free expansion slot of the right type.
- Host adapter: A host adapter normally supports some types of streaming interface, such as Serial Advanced Technology Attachment (SATA). Even though most people associate streaming interfaces with storage technologies, they can support other add-on types. As with any storage technology, make sure you have the appropriate cables when working with a device that uses a streaming interface. In addition, some devices will require an external, rather than an internal, host adapter connection, which usually means getting an add-on board in addition to the device.

Considering the Role of Software

Whenever you perform an expansion or update of your system, you need to consider the software that supports the hardware. Without appropriate software, the hardware might register, but you won't be able to use it. Expansion devices tend to cause more software problems because they're unique in functionality or expand the system in unexpected ways. You need to consider the effect of any change you make at these levels:

- Operating system: Older operating systems tend not to handle new devices very well. Make sure your new device is designed to work with the target operating system.
- Device driver: A device driver provides an interface between the hardware and the operating system. If you don't have a device driver that is designed to access the hardware fully, yet interact with the operating system on a machine, you'll discover that the device will work poorly—assuming it works at all.
- Configuration: Most devices require configuration to make them work optimally in a particular environment. All too often a vendor ships a device that lacks the appropriate configuration software for a particular operating system. You may need to rely on third-party alternatives to configure your device. Take time to research configuration needs in advance.

- **Application:** All of the installation and configuration tasks you perform aren't worthwhile if none of your applications actually work with the device in question. The problem can become quite serious at times. For example, you might think that a display adapter wouldn't have any problems with applications, but vendors often provide access to special versions of applications to ensure you can actually see the special features that the display adapter provides (the example system came with such special applications for this very reason).
- **Interaction:** Some devices are designed to interact with the outside world in some way. In this case, you need something more than a device driver, configuration software, or application to get the job done. You may need special operating system additions that make it possible for the device to access the outside world in a particular way, such as using a service or an agent to work with tape devices. The hardware vendor usually meets this particular need, so it's important to review all of the items provided with the hardware package and on the vendor site to ensure you get everything the device has to offer.

Obtaining and Installing Hardware Updates

Your hardware relies on firmware, software that resides in chips on the device, to accomplish tasks. The firmware provides instructions that tell the hardware how to interact with device drivers and how to perform basic tasks. As with any software, the firmware can get outdated, so you need to check things like BIOS revisions from time to time. The hardware vendor will provide the update software and instructions required to perform the firmware update. Regular updates will make the hardware more reliable, faster in some cases, and sometimes more secure as well.

> ### *WARNING*
>
> *Before you install any sort of hardware update, make sure you have a complete backup of your system data. You never know when an update will go wrong and you might lose access to your system until the vendor can create a fix for you. Take your time; read the instructions until you're certain you understand them before you perform the update.*

Some hardware rely heavily on drivers and services. If this is the case with your hardware, you'll also find that the services will check for updates automatically. Make sure you install these kinds of updates as quickly as possible to ensure your hardware remains reliable and provides full access to all the features it possesses. However, as with any update, make sure you have a backup of your system before you apply the update.

Don't assume that your hardware will check for updates. The vendor manual will often provide information on how to provide updates online. Make sure you visit online locations at least once a quarter (every 3 months) to check for updates that hardware services haven't performed automatically.

Knowing When to Retire Your Old System

Upgrading a system can become habit forming. One system I still own was upgraded 14 times before I finally called it quits and started over from scratch. During those upgrades, I added RAM, changed display adapters, upgraded the power supply, added new storage options, changed output ports, and so on. However, it finally got to the point that the case was frayed about the edges, the motherboard couldn't do the job, and the system as a whole was just plain worn out. It finally made sense to create a side table out of the thing and start over again.

> **NOTE**
>
> *All of my full tower case systems have become side tables over the years. It's helpful to replace the storage devices with drawers and the case itself is a nice conversation piece. I find the cases are just the right height for holding drinks and small lamps. Sometimes I tell people about the history behind a particular carcass. Theoretically, I could add lighting to the cases to dress them up by attaching lights of the correct wattage to the power supply outputs (I haven't actually tried it yet, but would love to hear about your experiences at John@JohnMuellerBooks.com). The old systems reside in my family room, where they remain quite happy.*

The trick is to know precisely when an old PC really can't be upgraded any longer. Here are some guidelines you can use to determine when your old system really has had it and you need to start over:

- The case is starting to fall apart—screw holes are elongated and components no longer seat correctly.
- The motherboard and one other main component, such as the display adapter, both require updates at the same time.
- The case feels too small for the number of items you're trying to put in it.
- Both software and hardware are outdated to the point that upgrades become expensive.
- A significant case component, such as an external switch, fails to work as expected. In addition, you note that external lights are hard to see and the connectors have become loose.

Index

A

Accelerated processing units (APUs), 36–37
Adapters, for devices not supported by display adapter, 110
Add-on boards, 23–27, 278
   checking for damage, 246
   display adapters as, 23–25
   host adapters as, 26
   POST cards as, 27
   security device cards as, 26
   sound cards as, 25–26
   SSDs as, 27
Add-ons, special, 277–278
Advanced Technology eXtended (ATX) motherboard, 18–20
   types of, 19–20
Agents, checking, 268–269
Aluminum cases, 18
AMD processors, 36–38, 40
Android, 200
Antistatic bags, 69
   danger of opening, 57
   for reusable parts, 13
Applications, 225–236
   circumventing conflicts between, 211
   connectivity requirements of, 228–229
   device driver requirements of, 217–218
   finding missing components of, 234
   hardware updates and, 279
   installing, 230–233
   matching to system, 226
   mismatches and, 233–234

Applications (*Cont.*):
   operating system support for, 198
   peer help with, 234–235
   removing, 236, 262–263, 264
   resetting an installation and, 235–236
   reviews of, 229–230
   for security, 261
   unpacking, 232
   validating platform requirements for, 227
   verifying hardware requirements for, 226–229
   version information for, 228
   Web-based, 232–233
APUs (accelerated processing units), 36–37
Arch Linux, 202
Assembly, waiting until you have all parts before beginning, 57
ATX. *See* Advanced Technology eXtended (ATX) motherboard
Automatic configuration, for SATA drives, 124–125
Auxiliary devices, 135–147. *See also specific devices*
   choosing, 136–137
   compatibility problems with, 136
   number of, 146–147

B

Backing up:
   before installing hardware updates, 279
   before installing new operating system, 204

Backups:
    of data, 266–267
    restoring, 269–279
    storing off-site, 267
    system, 218–219
Ball bearings, 32
Bare-bones kits, 7–8
Bearings, fans and, 32–33
Bias, in reviews, 42, 43, 44, 229–230
Bluetooth connectors:
    networking standards for, 183
    for wireless keyboards, 140
Box cutters, caution regarding use of, 57
Broadband connections, 170
Browsers, locking down, 176
Budget, creating, 9–11

C

Cable Internet connections, 170
Cables, 29–30
    checking, 248
    data, 29–30
    keeping tidy, cooling and, 30–31
    power supply, 29
    quality of, 30
    vacuuming, 240
Case plugs, connecting, 92, 94
Cases:
    configuring, 81–82
    connecting features to motherboard, 91–94
    material used to construct, 18
    opening, 16–18, 242–244
    orientations of, 18
    reusing, 11
    size of, motherboards and, 20
    specifications for, 18
    wiping, 240–241
CCleaner, 265
CEB (SSI Compact Electronics Bay), 20
CentOS, 202
ChromeOS, 199–200
Cleaning:
    cases, 240–241
    compressed air for, 240, 242, 243, 244, 245
    filters, 245

Cleaning (*Cont.*):
    hard drives, 264–265
    increasing cfm and, 33
    inside the case, 242–245
    work area, 65
Cleaning devices, 247
Cleanup utilities, 265
Clients, for LAN adapters, 155
Cloud storage, 28–29
    for backups, 266–267
COM port, for mouse and trackball
        connection, 142
Communication channel, protecting, 189–190
Compact Electronics Bay (CEB), 20
Compatibility, 45–47
    bad parts and, 46–47
    dealing with issues of, 47
    feature extension problems, 46
    of input devices, 136
    standards adherence problems and, 45–46
Components. *See* Parts; Parts comparisons;
        Spare parts
Compressed air:
    for cleaning cables, 240
    for cleaning inside case, 242, 243, 244, 245
Computer toolkits, 61–63
Configuration:
    of alternative devices, 186–187
    of common wireless devices, 185–186
    of display adapters, 110
    hardware updates and, 278
    of LAN adapters on motherboard,
        156–158
    of MODEMs, 173
    of motherboard, 106–107
    of multiple LANs, 167–172
    RAID support and, 133
    of SATA drives, 124–125
    system, 157
    unconventional, of device drivers, 216
    wired connections for, 147
    with wired keyboards, 137
Connectivity:
    of device drivers, ensuring, 222
    external, 33–34
    required by applications, 228–229

Index

Cooling, 71-74. *See also* Fans
   liquid, for processors, 74
   standard, for processors, 73-74
   thermal paste and, 72-73
Cooling fins, 30
CrossFire setup, 98-99
   configuration needs of, 105-108
   power supply compatible with, 99, 100
Custom systems, reasons to build, 4-6
Customizable off-the-shelf systems, 6-7

D

Damage, checking for damage and, 241-242
Data, security for, 266-267
Data cables, 29-30
   for storage devices, 120-122
Debian, 202
Defragmenting hard drives, 264-265
Denial of service (DOS) attacks, 188
Desktop cases, 18
Device drivers, 207-224
   application requirements and, 217
   automatic access and, 212
   avoiding device conflicts and, 210-211
   checking, 268-269
   circumventing application conflicts and, 211
   downloading, 205
   ensuring connectivity, 222
   existing, removing, 219-220
   hardware updates and, 278
   incompatible with applications, 217-218
   installing, 218-221
   mismatches and, 224
   missing devices and, 212
   operating system drivers and, 211-212
   operating system support of, 196-197
   operating system to device connections and, 208-211
   overcoming problems specific to, 221-224
   synching to operating system release, 216
   temporarily disabling, 222-223
   unconventional configurations of, 216
   updated versions of, 221
   vendor, advantages of, 213-215

Device drivers (*Cont.*):
   vendor, keeping updated, 215-218
   verifying settings and, 223-224
Dialog boxes, importance of reading, 262
Dial-up Internet connections, 170
Digital subscriber line (DSL) connections, 169, 170
Digital Video Interface (DVI), 98
DIMMs (dual-inline memory modules), 79-80
Direct memory access (DMA), 209-210
Disabling device drivers, temporarily, 222-223
Disasters, overcoming, 267-270
Disk Operating System (DOS), 196
Display adapters, 23-25, 98
   adapters for devices not normally supported by, 110
   adding, 276-277
   back of, 99-101
   configuring the case and, 110
   connecting cables and, 104-105
   connections between, 108
   CrossFire or SLI compatible power supply and, 99
   CrossFire or SLI setup and, 98-99, 105-108
   getting television reception without, 105
   installing, 101-103
   number of, 106
   power supply connections for, 108
   substitute, 55-56
   upgrading, 275-276
DisplayPort display adapters, 110
Displays, cleaning, 241
DMA (direct memory access), 209-210
Documentation:
   advisability of having paper copy of, 12, 61, 204
   complete, importance of, 60-61
   of device driver settings, 218
   of device settings, 218
   for existing parts, obtaining, 12
   importance of reading, 230
   for off-the-shelf systems, 5
DOS (denial of service) attacks, 188
DOS (Disk Operating System), 196

Drive size adapters, 122–123
Drivers. *See* Device drivers
DSL (digital subscriber line) connections, 169, 170
Dual-inline memory modules (DIMMs), 79–80
DVD backups, 266
DVI (Digital Video Interface), 98

E

eComStation, 200
EEB (SSI Enterprise Electronics Bay), 21
Electrical outlets, for work area, 65
Electromagnetic interference (EMI):
    configuring TV tuners to avoid, 108
    lack of, with wired keyboards, 137
    number of wireless devices and, 146
EMI. *See* Electromagnetic interference (EMI)
Encrypting data, 267
Enterprise Electronics Bay (EEB), 21
Ergonomics, wired keyboards and, 137
Ethernet connections, 33
Expansion:
    of customizable off-the-shelf systems, 7
    of standard off-the-shelf systems, 4
Expansion boards. *See* Add-on boards
External connectivity, 33–34
    network cards and, 33
    port cards and, 34
External drives, 117, 129–130
External LAN solutions, 159–160
Extras:
    listing for purchase, 58–59
    shopping for, 59
    storing, 60

F

Fans, 23, 30, 31–33
    bearings and, 32–33
    clamping in place, 77, 78
    connected to motherboard, 86
    maintenance of, 248–249
    powering, 77, 79
    spare, advisability of keeping, 32
    testing, 249

Feature extensions, compatibility problems with, 46
Fedora, 202
Filters, cleaning, 245
Firewalls:
    hardware for, 164
    types of, 176–177
Flash drives. *See* USB storage
Flex ATX, 20
Floppy drives, 27
    data cables for, 121–122
Form factors, 117–118
FreeDOS, 200
Full ATX, 19
Full tower cases, 18

G

Ghostery, 53, 54
Goals, writing down, 8–9
GPT (GUID partition table), 164, 165
Graphics cards. *See* Display adapters
GUID partition table (GPT), 164, 165

H

Hackintosh, 203
Haiku, 199
Hard drives, 28
    cleaning, 264–265
    hot swap functionality and, 116
    reusing, 11
Hardware:
    caution regarding security features provided by, 164
    errant, testing, 269
    for firewalls, 164
    inspecting hardware and, 245–248
    operating system support for, 198
    protection based on, 176, 177
    required by applications, 226–229
Hardware maintenance, 239–253
    add-on boards and, 246
    cables and, 248
    checking for damage and, 241–242
    cleaning filters and, 245

Index

Hardware maintenance (*Cont.*):
    fans and, 248-249
    opening the case for, 242-244
    schedule for, 252-253
    spare parts and, 250-252
    spraying dust out for, 244-245
    storage devices and, 246-247
    vacuuming louvers and cables and, 240
    wiping the case and, 240-241
Hardware updates, 271-280
    adding display adapters and, 276-277
    increasing storage and, 277
    installing, 279-280
    obtaining, 279
    for off-the-shelf PCs, 271-272
    processor plan and, 273-274
    software considerations and, 278-279
    special add-ons and, 277-278
    upgrading display adapters and, 275-276
    upgrading RAM and, 275
    wish list for, 272-273
HDMI (High-Definition Multimedia Interface), 98
High-Definition Multimedia Interface (HDMI), 98
Host adapters, 26, 278
Hot swap functionality, 116
Hotspots, 171
Hybrid hard drive and SSD, 28

I

IDE (Integrated Drive Electronics) data cables, 121
Identity theft, 188
Information Technology eXternded (ITX) motherboard, 19
Infrared connections, 161
Initial setup, testing, 95
Input devices. *See also specific devices*
    choosing, 136-137
Installation screens, importance of reading, 261-262
Installing:
    applications, 230-233
    device drivers, 218-221

Installing (*Cont.*):
    hardware updates, 279-280
    motherboards. *See* Installing motherboards
    operating systems, 205-206
Installing motherboards, 81-95
    configuring the case and, 81-82
    connecting case features and, 91-94
    identifying power plugs and sockets and, 84-85
    installing a power supply and, 82-83
    placing the motherboard and, 87-91
    positioning the motherboard and, 89-90
    screwing the motherboard in place and, 90-91
    setting up the motherboard and, 86-87
    standoffs and, 87-89
    testing the initial setup and, 95
Integrated Drive Electronics (IDE) data cables, 121
Integrated Services Digital Network (ISDN) connections, 171
Intel processors, 40
Interaction, hardware updates and, 279
Internal drives, 117
Internet connections, 167-177
    alternative options for, 174-175
    configuring multiple LANs and, 167-172
    MODEMs and, 172-174
    security and. *See* Internet security
    types of, 169-172
Internet security, 175-177
    firewalls and, 176-177
    locking down your browser and, 176
    locking down your system and, 176
    risk and, 175-176
    updates and, 175
Interrupt request (IRQ) settings, 209
ISDN (Integrated Services Digital Network) connections, 171
ITX (Information Technology eXternded) motherboard, 19

J

Jumper cables, 107, 108, 156-157

K

Keyboard connectors, 138, 139
Keyboard/mouse ports, 34
Keyboards, 136–140
    choosing, 136–137, 138
    wired, 137–139
    wireless, 139–140

L

LAN adapters, 152–153
    external, 159–160
    on motherboard, configuring, 156–158
    security and, 164
    separate, 158–159
    wireless, configuring, 185–186
LAN security, 163–166
    developing plan for, 163–165
    passwords and, 165
    with WLANs, 166
LANs. *See* Local area networks (LANs)
Laser connections, 161
Licensing, of operating systems, 203–204
Light, for work area, 64
Linux, 196, 197
    installing, 201–202
Linux Mint, 202
Liquid cooling systems, 31, 74
Local area networks (LANs), 151–166
    accidental association and, 188
    ad hoc, 188
    configuring adapter on motherboard and, 156–158
    connecting to display adapter, 169
    hardware for. *See* LAN adapters
    multiple, configuring, 167–172
    physical connection for, 153–154
    security and. *See* LAN security
    seeing other connections and, 155–156
    software for, 154–155
    wireless, 160–162
Louvers, vacuuming, 240
Low-level resources, device drivers and, 208–210

M

Mac OS X, 196, 197–198
    hacking, 203
MAC spoofing, 188
Mageia, 202
Maintainability, of off-the-shelf systems, 5
Maintenance. *See* Hardware maintenance; Software maintenance; Updates
Man in the middle, 188–189
Manuals, for operating systems, 204
Master boot record (MBR), 164, 165
MBR (master boot record), 164, 165
Mean time between failures (MTBF), 42, 250
MEB (SSI Midrange Electronics Bay), 21
Mesh network, networking standards for, 184
Micro ATX, 20
Micro SATA power cables, 120
Mid tower cases, 18
Midrange Electronics Bay (MEB), 21
Mini ATX, 19–20
Mini tower cases, 18
Mismatches, applications and, 233–234
Mobile Broadband Wireless Access, networking standards for, 184
Mobile Internet connections, 171
MODEMs, 172–174
    checking connection with a test system, 174
    configuring, 173
    connecting to, 174
    correct, obtaining, 172–173
    purpose of, 172
Monitors:
    cleaning, 241
    connecting to display adapter, 109
Motherboard:
    configuring the case and, 106–107
    fans connected to, 86
    identifying pins on, 92–93
    installing. *See* Installing motherboards
    multiple-LAN, 157–158
    PCIe slots and, 98–99
    role of, 18–21
    setting up, 86–87
    static electricity and, 87

Mouse, 140–141
　alternatives to, 141
　ports for, 140–141
　resolution of, 142
　wired, 142
　wireless, 143
　working without, 141
MTBF (mean time between failures), 42, 250

N

Network cards, 33
Networks:
　LANs. *See* LAN adapters; LAN security;
　　Local area networks (LANs)
　nontraditional, 188–189
　protecting, 190–191
Nontraditional networks, 188–189
　security and, 189
Non-writeable replaceable media, 117

O

Odors, maintenance and, 242
Off-the-shelf systems:
　customizable, 6–7
　drawbacks of, 3–4
　standard, 4
openSUSE, 202
Operating systems, 195–206
　alternative solutions for, 199–200
　backing up before installing, 204
　choosing, 195–200
　cleaning up the operating system and, 265
　device drivers specific to, 211–212
　DOS, 196
　downloading, 203–204
　downloading device drivers and, 205
　free, 203–204
　Hackintosh, 203
　hardware updates and, 278
　installing, 205–206
　licensed copy of, 203–204
　Linux, 196, 197, 201–202
　longevity of, 198–199
　Mac OS X, 196, 197–198, 203

Operating systems (*Cont.*):
　obtaining manuals for, 204
　security issues specific to, 189
　synching device drivers to release
　　of, 216
　vendor installation instructions for,
　　200–203
　version issues with, 197–198
　Windows, 196, 201
Optical drives, 28
　installing, 126
OS X. *See* Mac OS X

P

Parallel ports, 34
　for printers, 145–146
Parts:
　bad, working through, 46–47
　checking against list in box, 57
　choosing, 6. *See also* Parts comparisons
　comparison shopping for, 51
　compatibility pitfalls and, 45–47
　existing. *See* Reusable parts
　failure of, 46–47
　feature extensions and, 46
　list of, creating, 250
　order to follow in selecting,
　　38–40
　plastic, caution regarding, 82
　purchasing. *See* Purchase lists; Purchasing
　　parts
　refurbished, 10
　return policy for, 47
　reusable, 11–13
　reviews of, 42–46
　spare. *See* Spare parts
　standards adherence and, 45–46
　substitute, 55–56
Parts comparisons, 40–42
　performing efficiently, 1–42
　quick, on vendor sites, 37–38
　value of testing and, 40–41
Passphrases, 260
Password vaults, 260
Passwords, 165

PCIe. *See* Peripheral Component Interconnect Express (PCIe) slots
Peer help, 234–235
Pen tablets, 141
Performance, of off-the-shelf systems, 5
Peripheral Component Interconnect Express (PCIe) slots, 98–99
   correct, choosing, 107
   LAN adapters and, 153
Permanent storage, 115–133
   data cables for, 120–122
   drive size adapters and, 122–123
   external drives for, 129–130
   form factors and, 117–118
   hot swap functionality and, 116
   optical drives for, 126
   power cables for, 118–120
   RAID and, 131–133
   SATA drives for, 123–125
   SCSI option for, 130–131
   SSDs for, 115, 127–128
   USB, 128–129
Pixels, 98
Plastic cases, 18
Plastic parts, caution regarding, 82
Platform requirements, of applications, 227
Points, in reviews, 43
Ports, 34
POST (power on startup test) cards, 27
Power cables:
   for display adapter, 104
   for storage devices, 118–120
Power on startup test (POST) cards, 27
Power plugs, 84
Power supplies, 21–23
   calculating requirements for, 39–40
   connections for display adapters and, 108
   CrossFire or SLI compatible, 99, 100
   installing, 82–83
   quality of, 21–22
   reusing, 11
Power supply cables, 29
Power supply sockets, 84–85
Power supply Ys, 29, 30, 118–119
Printers, 145–146

Processors:
   AMD processors, 36–38, 40
   caution about pins on, 74
   cooling and. *See* Cooling; Cooling fins; Fans
   inserting, 75–79
   Intel, 40
   locking down, 76–79
   orienting, 75–76
   overclocking, 74
   planning for, 273–274
   testing, 40–41, 274
   verifying position for, 70–71
Projection systems, connecting to display adapter, 110
Protocols, for LAN adapters, 155
Puppy Linux, 202
Purchase lists, 50–52
   comparison shopping and, 51
   return policy and, 52
   specificity of, 50
   store reputation and, 50–51
   warranties and, 51–52
Purchasing parts, 52–56
   allowing scripts and, 53
   checking parts against list and, 55–56
   listing components and, 53–55
   verifying package contents and, 56–59

R

Radio frequency (RF) emissions, case material and, 18
RAID. *See* Redundant array of inexpensive disks (RAID)
RAM:
   inserting and securing, 79–80
   installing, 79–80
   upgrading, 275
   verifying position for, 70–71
RAM sockets, 79
ReactOS, 199
Read-write replaceable media, 118
Recording devices, connecting to display adapter, 109
Red Hat Enterprise Linux (RHEL), 202

Index

Redundant array of inexpensive
    disks (RAID):
  configuring the case and, 133
  levels of, 132
Refurbished parts, 10
Reliability:
  of off-the-shelf systems, 5
  of vendor device drivers, 215
  of wired keyboards, 137
Removing applications, 236, 262–263, 264
Restocking fees, 52
Restoring backups, 269–279
Retiring systems, 280
Return policies, 52
Reusable parts, 11–13
  antistatic bags for, 13
  obtaining documentation for, 12
  software required for, 13
  testing, 12
Reviews:
  of applications, 229–230
  checklist for, 44–45
  points in, 43–44
  reliable, sources for, 43
  standards adherence and, 45
RF (radio frequency) emissions, case material
    and, 18
RHEL (Red Hat Enterprise Linux), 202
Rollermouse, 141
Routers, 168–169

S

SATA. *See* Serial Advanced Technology
    Attachment (SATA) standard
Satellite Internet connections, 169, 171
Scalable Link Interface (SLI) configuration, 21
Scalable Link Interface (SLI) setup,
    98–99
  configuration needs of, 105–108
  power supply compatible with, 99, 100
Screws:
  overtightening, 91
  types and purposes of, 82, 83
Scripts, allowing, problems with, 53
SCSI drives, 130–131

Security, 259–263
  background applications for, 260–261
  Internet. *See* Internet security
  with LANs. *See* LAN security
  with nontraditional networks, 189
  operating systems and, 198
  passwords for, 260
  reading installation and update screens
      and, 261–263
  updates for, 261
  vigilance for, 260
  with Web-based applications, 233
  with wireless devices. *See* Wireless device
      security
Security device cards, 26
Self-contained devices, 117
Serial Advanced Technology Attachment
    (SATA) standard:
  advantages of, 130
  configuring the case and, 124–125
  data cables for, 120, 122
  installing drives and, 123–125
  power cables and, 119–120
Serial ports, 34
Server System Infrastructure (SSI)
    motherboard, 20–21
Services:
  checking, 268–269
  for LAN adapters, 155
Setup screens, importance of reading, 261–262
SFF (small form factor) cases, 18
Single-inline memory modules (SIMMs),
    79–80
Slackware Linux, 202
Sleeve bearings, 32–33
SLI. *See* Scalable Link Interface (SLI)
    configuration; Scalable Link Interface
    (SLI) setup
Slimline SATA power cables, 119–120
Small form factor (SFF) cases, 18
Software. *See also* Applications
  broken, 256
  defective, 52
  hardware updates and, 278–279
  for LAN adapters, 154–155

Software (*Cont.*):
  longevity of, 198–199
  maintenance of. *See* Software maintenance
  for reusable parts, 13
  security and, 261
  setting to check updates automatically, 256–257
Software maintenance, 255–270
  encrypting data and, 267
  overcoming disasters and, 267–270
  performing backups, 266–267
  required updates for, 255–259
  security and, 259–263
  system slowdowns and, 263–265
Solid-state drives (SSDs), 28, 115, 127–128
  as add-on boards, 27
  advantages and limitations of, 127
Sound cards, 25–26
Spare parts, 250–252
  advisability of keeping, 32, 252
  older, shopping for, 251
  parts list and, 250
  updated, 251
  using old parts as, 252
Specialty devices, reusing, 12
Speed, ports and, 34
Splitters, 29, 30, 118–119
SSDs. *See* Solid-state drives (SSDs)
SSI (Server System Infrastructure) motherboard, 20–21
SSI Compact Electronics Bay (CEB), 20
SSI Enterprise Electronics Bay (EEB), 21
SSI Midrange Electronics Bay (MEB), 21
Standard SATA power cables, 119
Standards:
  problems with adherence to, 45–46
  SATA. *See* Serial Advanced Technology Attachment (SATA) standard
Standoffs, installing, 87–89
Static electricity, 69–70
  antistatic bags and, 13, 57, 69
  discharging, 69–70, 87, 244, 274
  display adapter installation and, 107
  touching processor pins and, 74
SteamOS, 199

Storage devices, 27–29, 128–129
  checking, 246–247
  increasing, 277
Storage media. *See also* Permanent storage
  form factors and, 117–118
  options for, 117–118
Stores, reputation of, 50–51
SUSE Linux Enterprise, 202
Syllable, 200
System backups, 218–219
System configuration, 157
System reboots, when installing device drivers, 220
System slowdowns, 263–265
  cleaning the hard drive and, 264–265
  cleaning up the operating system and, 265
  removing unused applications and, 264

T

Tape backups, 266
Tape drives, 28
Television tuners, configuring the case and, 108–109
Televisions:
  connecting to display adapter, 109
  getting reception without a special display adapter, 105
Testing:
  after operating system installation, 169
  of basic setup, 144–145
  of errant hardware, 269
  of fans, 249
  of initial setup, 95
  of MODEM connection, 173
  of processors, 40–41, 274
  of updates, 259
  of video setup, 112–113
Thermal paste, 72–73
Thumb drives. *See* USB storage
Tools:
  computer toolkit and, 61–63
  individual, 63–64
  magnetic, caution regarding, 61
  required, 61

Index

Tower cases, 18
Trackballs, 141–143
   resolution of, 142
   wired, 142
   wireless, 143

U

Ubuntu, 202
UEFI (Unified Extensible Firmware Interface), 164
Ultra wideband (UWB), networking standards for, 183
Unified Extensible Firmware Interface (UEFI), 164
Uninstall applications, 264
Universal Serial Bus (USB). *See entries beginning with term* USB
Updates:
   for device drivers, 215
   downloading and installing, 258–259
   Internet security and, 175
   for security software, 261
   setting software to check updates automatically and, 256–257
   software, required, 255–259
   testing, 259
   validating platform requirements for, 257–258
USB (Universal Serial Bus) port plug-ins, 277
USB (Universal Serial Bus) ports, 34
   for mouse and trackball connections, 142
   for printers, 145
USB connectors, for wireless keyboards, 140
USB external data cables, 120–121
USB internal data cables, 121
USB storage, 28, 128–129
UWB (ultra wideband), networking standards for, 183

V

Vacuum cleaners, caution regarding magnetic field of, 240
Vacuuming louvers and cables, 240

Vendor sites, 36–40
   order to follow in visiting, 38–40
   quick comparisons using, 37–38
Vendors:
   device drivers from, 212–218
   information provided by, 47–48
   operating system installation instructions from, 200–203
   reputation of, 50–51
Video, 97–113
   connecting cables and, 104–105
   connecting devices and, 109–111
   CrossFire and SLI configuration needs for, 105–108
   display adapters and. *See* Display adapters
   PCIe slots and, 98–99
   power supply for, 99, 100
   quick test of, 112–113
   TV tuner configuration needs and, 108–109
Video cards. *See* Display adapters
Video port plug-ins, 277–278
Viruses, software malfunctions and, 256

W

WAPs (wireless access points), 191
Warranties:
   reading carefully, 51–52
   on refurbished parts, 10
Web-based applications, 232–233
Webcams, 146
WEP (Wireless Equivalent Privacy), 189–190
WIDS (Wireless Intrusion Detection System), 191
Wi-Fi. *See* Wireless fidelity (Wi-Fi)
Wi-Fi Protected Access (WPA) Level 1, 190
Wi-Fi Protected Access Level 2 (WPA2), 190
Wi-Fi Protected Setup, 190
WiMAX (Worldwide Interoperability for Microwave Access) devices, 162
Windows, 196
   installing, 201
   restore point technology of, 270
WIPS (Wireless Intrusion Prevention System), 191
Wired connections, for LANs, 153

Wired keyboards, 137–139
Wired mouse, 142
Wireless access points (WAPs), 191
Wireless connections, 179–191
    alternatives for, 161
    common wireless standards and, 180, 181–184
    for LANs, 154
    standards support and, 180, 185
    transmission speeds and, 180
Wireless device security, 187–191
    intrusion types and, 188–189
    need for proactive approach to, 187
    protecting communication channel and, 189–190
    protecting the network and, 190–191
    WAPs and, 191
Wireless devices:
    configuring the case and, 185–186
    security and. *See* Wireless device security
Wireless Equivalent Privacy (WEP), 189–190
Wireless fidelity (Wi-Fi), 161–162
    networking standards for, 181–183
Wireless Intrusion Detection System (WIDS), 191
Wireless Intrusion Prevention System (WIPS), 191
Wireless keyboards, 139–140

Wireless local area networks (WLANs), 160–162
    defining, 160–161
    security and, 166
    Wi-Fi and, 161–162
    WiMAX and, 162
Wireless Metropolitan Area Networks (WMANs), networking standards for, 184
Wireless mouse, 143–144
Wireless personal area networks, networking standards for, 183
Wireless receivers, generic vs. special-purpose, 143–144
Wireless Regional Area Networks, networking standards for, 184
Wireless trackballs, 143–144
WLANs. *See* Wireless local area networks (WLANs)
WMANs (Wireless Metropolitan Area Networks), networking standards for, 184
Work area, 64–65
Work table, 64
Worldwide Interoperability for Microwave Access (WiMAX) devices, 162
WPA (Wi-Fi Protected Access) Level 1, 190
Wristbands, to discharge static electricity, 70
Write-once replaceable media, 117–118

Z

ZigBee, networking standards for, 183
ZIP drives, 27